CW01302151

OFF THE RAILS

OFF THE RAILS

The Inside Story of HS2

Sally Gimson

ONEWORLD

A Oneworld Book

First published by Oneworld Publications Ltd in 2025

Copyright © Sally Gimson, 2025

The moral right of Sally Gimson to be identified as the Author of this work has been asserted by her in accordance with the Copyright, Designs, and Patents Act 1988

All rights reserved
Copyright under Berne Convention
A CIP record for this title is available from the British Library

ISBN 978-1-83643-017-9
eISBN 978-1-83643-018-6

Map by John Gilkes
Typeset by Geethik technologies
Printed and bound in Great Britain by Clays Ltd, Elcograf S.p.A.

Oneworld Publications Ltd
10 Bloomsbury Street
London WC1B 3SR
England

No part of this publication may be reproduced, stored in a retrieval system, or transmitted, in any form or by any means, electronic, mechanical, photocopying, recording or otherwise, or used in any manner for the purpose of training artificial intelligence technologies or systems, without the prior permission of the publishers.

The authorised representative in the EEA is eucomply OÜ,
Pärnu mnt 139b–14, 11317 Tallinn, Estonia
(email: hello@eucompliancepartner.com / phone: +33757690241)

Stay up to date with the latest books, special offers, and exclusive content from Oneworld with our newsletter

Sign up on our website
oneworld-publications.com

MIX
Paper | Supporting responsible forestry
FSC® C018072

Contents

Map		ix
Introduction		1
1	A Very British Project	13
2	The Rise of London and the Motorcar	33
3	A National Mission	47
4	In the Public Interest	60
5	Rebellion in the Shires	79
6	Clearing a Path Through the City	89
7	*Grand Projet* Angst	98
8	McLoughlin to the Rescue	106
9	Obama Tactics	117
10	The Naysayers	131
11	Hybrid Wars	140
12	Euston, We Have a Problem	156

13	Another Rethink?	164
14	The Special Minister for HS2	175
15	A Green Revolution	193
16	'Robbing the White Elephant to Pay the Red Wall'	204
17	Our Friends in the North	221
18	In the Path of the Ghost Train	234
19	A Disaster for Crewe	240
Conclusion		251

Acknowledgements	265
Glossary	267
Further Reading, Viewing and Listening	273
Index	279

For Eliza, Clive and Katy

HS2's Original Route

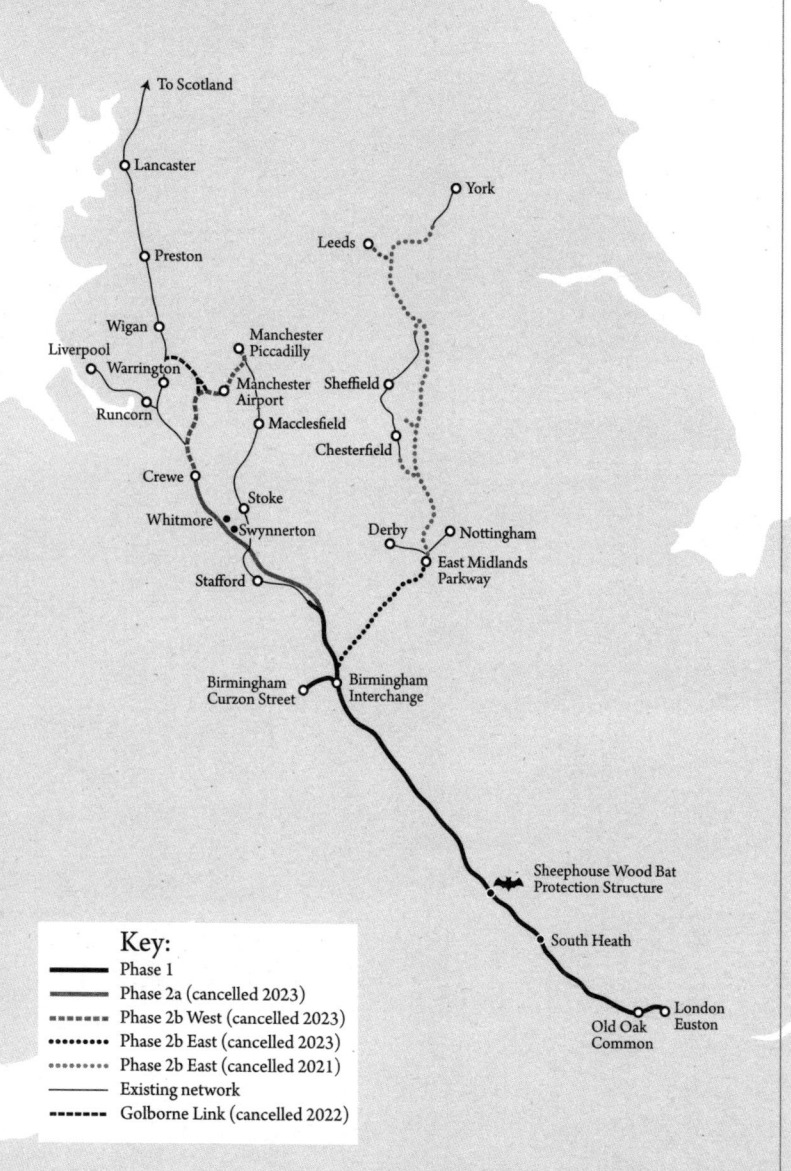

To Scotland
Lancaster
Preston
Wigan
Liverpool
Warrington
Runcorn
Manchester Piccadilly
Manchester Airport
Macclesfield
York
Leeds
Sheffield
Chesterfield
Crewe
Whitmore
Stoke
Swynnerton
Stafford
Derby
Nottingham
East Midlands Parkway
Birmingham Curzon Street
Birmingham Interchange
Sheephouse Wood Bat Protection Structure
South Heath
Old Oak Common
London Euston

Key:
- Phase 1
- Phase 2a (cancelled 2023)
- Phase 2b West (cancelled 2023)
- Phase 2b East (cancelled 2023)
- Phase 2b East (cancelled 2021)
- Existing network
- Golborne Link (cancelled 2022)

Introduction

Nasrine Djemai was twelve when she was informed that the home she had lived in all her life would be flattened. It was 2009. The world was gripped by a global recession, the deficit had quadrupled and unemployment had risen rapidly. A general election was on the horizon and more than a decade of New Labour rule was coming to an end. Families across the country were struggling to make ends meet and now Nasrine's was facing homelessness. The government had decided to drive a high-speed train line to Birmingham, Manchester and Leeds through her neighbourhood. At first, she couldn't believe it, but the next day she watched as her estate was invaded by men in hi-vis jackets marking the roads, taking photographs and setting up tripods. Nasrine's block was to be razed to build the approach of HS2 into Euston station.

Today, more than fifteen years later, Nasrine's neighbourhood remains a building site. She and her family have been re-housed, but the railway line into Euston has not been completed and plans for the station remain theoretical. No one is quite sure when, or if, HS2 will reach central London. There are no longer plans to run high-speed trains to the North, instead the new track will extend from the newly built station at Old Oak Common, a rather desolate neighbourhood in north-west London, to an even more obscure junction beyond Birmingham in the West Midlands. The vision for HS2 – once feted as Europe's largest infrastructure project, the first new railway built north of the capital since 1900, an opportunity to rebalance the London-centric economy by creating reliable connections to the North – is in tatters.

What should have been a source of national pride is now a multi-billion-pound bleeding stump of a line, a rebuke to the inability of successive, reckless Conservative governments to decide what kind of a country Britain should be. Even the most ardent advocates of high-speed rail now believe the HS2 line will be a white elephant unless it is brought into Euston and extended beyond Birmingham. The Labour government elected in July 2024 has pledged HS2 will eventually terminate at Euston, but ministers are unable to say when this will happen or how much it will cost and the project's chief executive has warned that things will get worse before they get better.

So, what went wrong?

Certainly, HS2's final death knell sounded in October 2023 when then prime minister Rishi Sunak stood up at a Conservative Party conference and announced, with little warning, the cancellation of the railway's northern leg. It would not be extending beyond Birmingham. Furthermore, neither Euston station nor its approaches would be funded without private investment. The announcement landed with a thud. Hundreds of millions of pounds had been spent on planning the approach into Euston and the line's northern half. Debates over HS2 had taken up more than 1,200 hours of parliamentary time. An act for the

INTRODUCTION

high-speed train to run between Birmingham and Crewe had been passed in parliament. For the prime minister to announce such a major scaling back unilaterally, without preparing MPs, was an insult to democracy. But Sunak was a desperate man. HS2, which had become a byword for financial mismanagement, was an easy target – a convenient source of cash to fill potholes in marginal constituencies.

While government interference was nothing new, there was outrage and disbelief in global engineering circles. How could a major part of such a huge infrastructure project – a new fast railway across a G7 country – be cancelled so abruptly? For international observers and investors, the announcement was yet another sign that Britain was not a serious place to do business. And all the rhetoric successive Tory prime ministers and chancellors had spouted over the 2010s about levelling up, the Northern Powerhouse or rebalancing the economy was revealed to be simply hot air. When the political and economic waters became rough, the Tory prime minister was quite happy to dump an unpopular project.

Fundamentally, high-speed rail is not a bad idea. Britain's Victorian – and in some cases Georgian – rail network is splitting at the seams. The unreliability and slowness of services up the west coast of Britain and across the country's northern belt from Manchester to Hull has long hampered growth outside London and the South East. Merely repairing and electrifying 200-year-old tracks and patching up old tunnels and bridges is increasingly expensive and inadequate, and does not solve the other fundamental problem – both the railways and roads in the North West are inadequate to accommodate the increasing movement of goods and people around the country.

Plans to replace some Victorian railways had been on the agenda throughout the 2000s, but the political appetite to invest in infrastructure outside London wasn't there and the Treasury and the then chancellor Gordon Brown were dead against what they considered to be a *grand projet*, as the French TGV was sneeringly described. The high-speed rail idea was endorsed with lukewarm enthusiasm in the 2005 Labour manifesto as something the government would investigate, but not necessarily pursue. While the city of London was powering the country and growth could be distributed in receipts around the country, there was little imperative to embark on what might prove to be a risky, new, fancy railway project towards the North.

After the financial crisis of 2008 that calculation changed. The new prime minister, Gordon Brown, had a Damascene conversion. He realised that if anything went wrong with London, there were no other cities or regions in the UK where growth was strong enough to take up the slack. As a recent analysis by the *FT* has shown, the UK is still an outlier among other wealthy European countries because of its dependence on the capital. Even in traditionally centralised France, the difference in wealth between Paris and Lyon is far smaller than that between London and Manchester. Take London out of the equation and the UK has the economy of Mississippi. Brown believed a large infrastructure project might shift the dial and lay the foundations to rebalance this wealth gap. Luckily for Brown his rail minister (and then transport secretary) Lord Andrew Adonis had a plan in his back pocket.

A train geek, Adonis had a vision for British high-speed rail and believed passionately that Britain could emulate France, Germany and Spain and reap the same economic benefits that

INTRODUCTION

connecting city regions had brought them. He wanted the whole of Britain – not just London – to look like its modern European counterparts. Instead of 'making do and mending', and relying on its past industrial glories, Adonis believed in the need to drag the country into the twenty-first century and viewed a British high-speed rail network as his own personal transformative legacy.

So far so good. The trouble was that by the time the Labour government decided to build a high-speed train network, with the Tories agitating belatedly for one too – the Labour government was approximately two years out from a general election it was likely to lose. That was a very short time in which to sell the idea of a new high-speed rail line to the general public, let alone do anything as radical as consult them on a route. So Adonis didn't. He set up a wholly owned government company called HS2 Ltd and asked a special group of civil servants and engineers to come up with a plan as quickly as possible to present to parliament before an election. The cost was an entirely notional £30 billion – an optimistic finger in the air estimate based on the cost of high-speed rail in Europe. The route was set by engineers, who tried to avoid as many houses as possible on their way to Birmingham and beyond and the specifications the highest possible so the trains could run fast and smoothly without disturbing anyone. There was enough information for a command paper to be written, but not much else before Labour was booted out in 2010 and the Conservative–Liberal Democrat coalition government under David Cameron took over.

But the command paper whetted the appetite of the next government. Despite viciously slashing money for public services, the new coalition ministers promised to invest in infrastructure. It was the run-up to the 2012 Olympics and the 'golden age' of British–Chinese relations. Britain was still a member of the EU and George Osborne, then chancellor of the exchequer, believed modernity meant iconic new stations and high-speed lines to northern cities. Politicians viewed HS1, the first high-speed line in Britain, as an enormous success – running 68 miles from London to the Channel Tunnel, carrying both the Eurostar and the domestic Javelin train to Kent, it had helped London win the 2012 Olympic Games bid as the final section could shuttle thousands of athletes and attendees from the Olympic Stadium in east London to St Pancras International.

But no large-scale project had been built north of London in modern times. HS2 was revolutionary in its stated intent of joining up Britain though a modern public transport system and providing balance to London's financial monopoly. Theoretically, it was an example of 'cathedral thinking' – a rarity in politics. It would pioneer policy and engineering expertise for the sake of national socio-economic improvement to the whole country, largely for the benefit of future generations.

***

I wanted to write this book to understand why HS2 proved so very challenging and why the costs skyrocketed as they did. The reduced line today is likely to come in at the £80 billion mark (although no one is still quite sure), more than ten times as expensive as high-speed rail across the Channel. Like many others, I was agnostic about HS2 when I started researching. It seemed very expensive. I didn't really understand why it was being built, nor why it cost so much. And I too wondered whether it was worth doing. I also felt strongly that Britain's troubles with

INTRODUCTION

HS2 mirrored the rocky politics of the last fifteen years, which was further destabilised by the vote to leave the EU.

HS2 represented the pre-Brexit liberal view of the world embodied by the Olympics: modern, European, optimistic, driven by cities and a new industrial future for the North. After Brexit, HS2's purpose became confused as the politicians who had sponsored the railway were swept away and politics became a more parochial fight for the towns, suburbs and communities which HS2 bypassed (although it would undoubtedly have made them better off). Buffeted by Covid and the Ukraine war, afraid of an uncertain future which might further disrupt old assumptions, politicians – and voters, too – became less confident, searching for quick fixes and becoming even less willing to invest in the long term. Did the story of HS2 shed more light on what was going on socially and politically in Britain at the time? How on earth did Britain make such a mess of building a high-speed line when they are more or less the norm all over the world; in Spain and Japan high-speed rail projects engender national pride – why then did HS2 attract so much scorn here?

And it was people who love British trains, seemingly natural allies of an exciting new rail line, who became the project's greatest detractors – perhaps because HS2 threatened to disrupt the status quo and challenge old myths about the great age of British railways. HS2 was an admission that the old railway system, built in the glory days of the British Empire, was no longer fit for purpose. For people who had spent their whole lives treasuring the past, this was identity-threatening stuff. It meant Victorian lines might become redundant, old tunnels changed, the trains they loved put out of service. The past, rather than a bygone idyll, might be found wanting.

This book will also show what a hugely complex endeavour HS2 was, far more than HS1 (the high-speed rail link from the Channel Tunnel to St Pancras) or Crossrail (the Elizabeth line) to which it is often compared. It is still the largest infrastructure project being built in Europe. Nothing on this scale had been tried before in modern Britain. And the men (mostly) in charge of commissioning and managing the project were hopelessly out of their depth.

This book will show the damage that lack of political clarity – and then constant political interference – does to a major infrastructure project. While MPs and ministers may blame the failure of HS2 on planning laws, 'not in my back yard' (NIMBY) environmental campaigners and overpriced bat-protection tunnels, I'll demonstrate that it was politicians and parts of the civil service who were most to blame: MPs sitting on parliamentary committees demanding unnecessary changes that would cost hundreds of millions; ministers failing to lay out clear objectives from the beginning and then changing them with the wind; prime ministers refusing to be tied down to a broader UK transport strategy and insisting on tunnelling or cuts to vital parts of the line to pay off opponents and a Treasury which continually tried to slash costs without any understanding of modern civil engineering.

Meanwhile those who understood how to deliver complex global projects and believed in the enormous benefits that HS2 would bring found themselves snarled up in toxic politics and constant scrutiny from hostile opponents and eventually walked away. The large contractors shrugged and added another zero to their bill, although they too became increasingly frustrated. This book will also look at why environmentalists failed to support

INTRODUCTION

what was ultimately a green transport project and campaigned against HS2 with as much energy as they had against nuclear power stations or the third runway at Heathrow.

***

I come from Scotland. Connections and infrastructure matter to me. They shape how a country sees itself and its place in the world. The lives of my cousins on Skye were transformed by a simple bridge which connected them to the mainland. The advent of the high-speed 125 train in 1981 and the electrification of the line to Edinburgh changed my life. University in England was suddenly an option for my sisters and me. Transport links *matter*.

Between 2011 and 2018, I was also a Labour councillor in Camden. I witnessed first hand the extraordinary difficulties of bringing a train into a heavily built-up part of London, the harm to the lives of our least well-off residents who would be forced to live on a building site for up to twenty-five years, the lack of leadership and the inability of all parties involved to work for the national good and the mixed messaging about what HS2 was for. I was also briefly the parliamentary candidate in Bassetlaw in North Nottinghamshire, before being ousted for a more Corbyn-sympathetic candidate, and met communities who desperately needed new infrastructure investment of all kinds to thrive.

While HS2 would only have provided a fast connection between Sheffield and Leeds, there would have been a knock-on effect for towns like Worksop. I want to understand why the importance of transport connection was lost and why HS2 was considered a stand-alone project, never embedded in a 'national transport plan'. Even the idea that HS2 had a vital role to play in the North's economies was underplayed, with only experts really comprehending the larger issue of congestion on the West Coast Main Line. HS2 would free up capacity for freight and stopping trains, acting as a kind of fast bypass to Scotland, meaning *non*-HS2 services would also run more reliably and many towns would be much better served. Was this because all the commentators lived in London?

Even as this 'green project' was being discussed there was a refusal to talk about how to prevent traffic on roads and reduce air travel. Was this deliberate, born out of some fear of upsetting people? Or just an ideological refusal to reduce 'choice'? After all, France has tolls on motorways and China and Japan have more or less stopped air travel between some major cities because of high-speed rail. And why was there no discussion of ticket prices or who the train was for? Was this the consequence of a top-down project? The vacuum had consequences. One civil servant said when he went to talk to a leading London girls' school about HS2, the students told him it was a luxury train for rich people. In vain he tried to persuade them it could be a 'people's train'.

There are countless reasons HS2 floundered, each uniquely revealing of British society today. This is a tale of civil servants expert at writing policy papers, poring over spreadsheets, drafting legislation and steering laws through parliament, but with no engineering experience, who thought that they could 'learn on the job' while commissioning and overseeing a high-speed megaproject the size of which had never been attempted anywhere else in the world. It is also a tale of government mismanagement, stringent planning laws and political interference; an unwillingness to engage with 'experts' and the true cost of austerity. It's a

INTRODUCTION

story of how top-down projects are no longer possible in the age of online warriors and populism. There are huge lessons to be learned from HS2 for the new Labour government intent on growth and who will need to deal with the world as it is now. Just imposing infrastructure on the country will not be enough. Hopefully, this book will help MPs reflect on why.

***

To write this book, I've talked to scores of people up and down the line: government ministers who made some of the key decisions, including Lord Patrick McLoughlin, the secretary of state for transport under David Cameron, and Andrew Stephenson, the first minister for HS2; technical experts like Andrew McNaughton, the engineer who designed HS2, and Allan Cook, the engineer chair of the HS2 who ensured building started; senior civil servants from the Department for Transport (DfT), as well as environmentalists worried about the effect on Britain's countryside.

I spoke to Andy Street, former mayor of the West Midlands, now Lord Street, whose vision for the region relied on high-speed rail, and the mayor of West Yorkshire, Tracy Brabin, who discovered the line to her city of Leeds was cancelled only months after she was elected. I also travelled along the line and interviewed people who lost their homes and those left worried about HS2's effect on their villages. Each had a different tale to tell. Some of my interviewees came to be great supporters of HS2, while others hardened their opposition over the years. I've read endless reports about HS2 – produced with alarming regularity by HS2 Ltd itself, parliamentary committees, quangos, thinktanks, government departments – each with a slightly different point of view. So many, in fact, that it makes your head spin.

When you have finished this book, I hope you will understand why HS2 was so complicated, cost so much and why it was constantly derailed. I hope you will read enough evidence to decide for yourself whether it is worth investing billions of pounds of public money in future high-speed rail. But, most importantly, I hope that you might consider how we can build a better political system to stop such an expensive and embarrassing fiasco ever happening again.

If we ask another twelve-year-old like Nasrine to give up her home, let's at least make sure it's worth it.

1

A Very British Project

HS2 could have been a thrilling project, sparking the collective imagination. A high-speed line for a high-speed society; Birmingham businesspeople would hold meetings with their colleagues in sleek first-class carriages, transforming the Northern Belt and the Midlands into a new industrial powerhouse to rival London. Schools and tour groups would flock to Manchester's art galleries and museums. Travellers across the UK would now have a fast route to Manchester and Birmingham airports, reducing the need for a third runway at Heathrow. A modern-day Hogwarts Express, families could share sandwiches while couples sipped champagne in the dining car. Yorkshire's tourism could take off, drawing international visitors from London, while Birmingham's picturesque canals might prove catnip for day trippers. Showing off Britain's biggest cities to the world, HS2 would introduce the Midlands and the North to the global stage as places to invest in and explore.

Of course, Britain has always been proud of its railways, because we invented them. When the great railway engineer Robert Stephenson died in 1859, he was buried in the nave of Westminster

Abbey, laid alongside Sir Isaac Newton, the playwright Ben Jonson and the most celebrated Scottish surgeon and anatomist of the eighteenth century, John Hunter. His magnificent funeral hearse, drawn by six horses, travelled from his home in Gloucester Square through Piccadilly and across Hyde Park, for which special permission had to be sought from Queen Victoria. People lined the route of the funeral cortège and between three and four thousand crowded into the Abbey for the service. In the towns and cities of his native Northumbria, as well as in Whitby, where he had served as an MP for over a decade, businesses closed and ships flew their flags at half-mast. Church bells were muffled in mourning.

Robert and his father George were pioneers in railway technology and many of the lines they built across Britain are still in use today. Between them, they designed the arched viaducts, elaborate tunnels and decorated ventilation shafts still embedded in the English countryside. And most importantly, they invented the iconic steam locomotives which ran on those lines, enabling the Industrial Revolution to take flight and lining the British Empire's pockets. This technology was soon sold across Europe and America, while the Empire was drawn tighter into Britain's sphere, with railways branching through India, Australia, New Zealand, Hong Kong and beyond. Samuel Smiles, the moralist of the day, devoted his book, *Lives of the Engineers*, to the Stephensons' endeavours. They were the Steve Jobs and Bill Gates of their age, proof that anyone in Victorian Britain with talent and the right moral fibre could become national heroes.

The elder, George Stephenson, had an unlikely start for a Victorian entrepreneur, initially working as a brakeman at Killingworth Colliery in Tyneside, responsible for the winding mechanism that lowered the cage of miners into the pit shaft. Like many of his peers, George was illiterate until the age of eighteen, however, his practical solutions to the mine's many technical problems were so ingenious that he quickly became the colliery's chief engineer. He drew up plans for steam locomotives – which initially looked like boilers on wheels with a chimney at the front – that could pull thirty tonnes of loaded coal wagons up a 1:450 slope, massively reducing the physical toll on the miners and horses. When the Quaker entrepreneur Edward Pease wanted to join the coal mines at Shildon in County Durham to Darlington, he approached George and his young son Robert to help. The two men convinced Pease that the eight-mile track he wanted to build could use steam engines instead of horses – and they built the Stockton to Darlington line with his backing.

The Stockton–Darlington route established the Stephensons' reputation immediately and the merchants, traders and manufacturers of Manchester and Liverpool seized on the railway as a potential solution to their problems. The new, mechanised factories of the North had found their growth stifled by the sluggishness of nineteenth-century transport: either a cart over a bumpy turnpike road or a slow and expensive canal journey. Both were costly, inefficient and vulnerable to poor weather (not unlike some cross-country services today). They were controlled by the gentry, over whose land roads passed and rivers ran. Not only could goods be moved a lot faster by train, but companies could design their railways to suit their needs. A twelve-hour canal journey between Manchester and Liverpool could be reduced to two hours, knocking all competition out of the park.

So, the merchants, traders and manufacturers banded together to form a railway company. However, because the line ran over private land, the company needed parliament's permission via a so-called private bill. The bill would grant them legal powers to start work while also allowing the landowners affected to petition for compensation or request changes to the route – a system very similar to that in place today, although there were far fewer landowners to square off in the nineteenth century.

When it was first introduced to parliament, the Manchester–Liverpool railway bid failed. Lords Derby and Sefton objected to the route passing through their estates, as did the Duke of Bridgewater who owned the canal and stood to lose substantial revenue. MPs cast doubts on the company's surveys, arguing that the whole project would cost closer to £200,000 than the proposed £40,000, mainly due to the difficulties of crossing a bog called Chat Moss. George Stephenson, brought in to defend his plans, was so tongue-tied in parliament when faced with the Cambridge-educated lawyer Edward Alderson that he couldn't answer any of the technical details. The bill failed. Later, he admitted: 'I was not long in the witness box before I began to wish for a hole to creep out at.'

Stephenson was sacked and another set of engineers employed. The route was altered and landowners paid off. Fortunately for Stephenson, the replacement engineers lacked his technical aptitude and he was soon rehired. When the Manchester to Liverpool railway was finally completed in 1830, costs had rocketed past MPs' estimates to £280,000. However, unlike HS2, George Stephenson and his assistant Thomas Gooch completed it in just four years – without the aid of modern technology. The final route consisted of sixty-three bridges including the Sankey Viaduct which spanned the Sankey Valley and canal, today a Grade I-listed monument – the oldest railway viaduct in the world, still in use after electrification in 2015. They tackled the enormous technical challenge of crossing the twelve-mile-squared peat bog at Chat Moss by floating the line over rafts made of bundles of heather, brushwood mattresses and panels woven from branches.

After defeating a clutch of lesser locomotives at the Rainhill Trials, *Rocket* – the steam engine built by George's son Robert – was chosen as the locomotive to pull passenger trains. Although the government had played little role in its construction, the line was opened by the ageing prime minister, the Duke of Wellington. The only glitch in the proceedings was the death of the MP William Huskisson, hit by *Rocket* after he alighted the train to discuss politics with Wellington – described contemporaneously by actress and abolitionist Fanny Kemble:

> The engine had stopped to take in a supply of water, and several of the gentlemen in the directors' carriage had jumped out to look about them. Lord W——, Count Batthyany, Count Matuscenitz, and Mr. Huskisson among the rest were standing talking in the middle of the road, when an engine on the other line, which was parading up and down merely to show its speed, was seen coming down upon them like lightning. The most active of those in peril sprang back into their seats… while poor Mr. Huskisson, less active from the effects of age and ill health, bewildered, too, by the frantic cries of 'Stop the engine! Clear the track!' that resounded on all sides, completely lost his head, looked helplessly to the right and left, and was instantaneously prostrated by the fatal machine, which dashed down like a thunderbolt upon him, and passed over his leg, smashing and mangling it in the most horrible way. (Lady W—— said she

distinctly heard the crushing of the bone.) So terrible was the effect of the appalling accident that, except that ghastly 'crushing' and poor Mrs. Huskisson's piercing shriek, not a sound was heard or a word uttered among the immediate spectators of the catastrophe. Lord W—— was the first to raise the poor sufferer, and calling to aid his surgical skill, which is considerable, he tied up the severed artery, and for a time, at least, prevented death by loss of blood. Mr. Huskisson was then placed in a carriage with his wife and Lord W——, and the engine, having been detached from the director's carriage, conveyed them to Manchester. So great was the shock produced upon the whole party by this event, that the Duke of Wellington declared his intention not to proceed, but to return immediately to Liverpool. However, upon its being represented to him that the whole population of Manchester had turned out to witness the procession, and that a disappointment might give rise to riots and disturbances, he consented to go on, and gloomily enough the rest of the journey was accomplished.

Despite this minor setback, the celebratory banquet continued as planned. And the Manchester to Liverpool line was a roaring success. The greatest surprise was how the trains captured the public imagination. Around 50,000 people – the equivalent of a third of the population of Manchester at that time – rode the railway in the first three months. Souvenirs and memorabilia were cast. Commemorative mugs, postcards, prints and handkerchiefs were sold depicting the engine, the line and the hundreds of passengers sitting in the open carriages. The *Penny Post* proclaimed how a 'pleasurable wonder takes possession of the mind, as we glide along at a speed equal to the gallop of a racehorse'.

The age of the railways had begun. Businessmen from the towns and cities which were growing around the country wanted a piece of the action and many banded together to set up railway companies. Those in the countryside were more cynical. George Eliot captures the mood in her novel *Middlemarch*:

> In the hundred to which Middlemarch belonged, railways were as exciting a topic as the Reform Bill or the imminent horrors of Cholera, and those who held the most decided views on the subject were women and landholders. Women both old and young regarded travelling by steam as presumptuous and dangerous, and argued against it by saying that nothing should induce them to get into a railway carriage; while proprietors...were yet unanimous in the opinion that in selling land, whether to the Enemy of mankind or to a company obliged to purchase, these pernicious agencies must be made to pay a very high price to landowners for permission to injure mankind.

***

Railway companies began building lines all over the country, north–south as well as east–west. In 1838, the London to Birmingham line, now part of the West Coast Main Line (which would be bypassed by the HS2 route), was completed to mark the coronation of the eighteen-year-old Queen Victoria. The railway cost £5.5 million (£702 million today) and took 20,000 men four years to construct. The railway company appointed Robert Stephenson as chief engineer. During the planning stages, he travelled the route the line would take, speaking personally to the landowners affected. Similar to the concessions made for landowners on the HS2 route, the earls of Essex and Clarendon forced Stephenson to divert around the Gade Valley, south of the Chiltern Hills, because it would have passed through Grove and Cassiobury parks.

In Northampton, Stephenson was compelled to design the mile-long Kilsby Tunnel, still in use today and restored as recently as 2020, to circumvent other country estates. At the time it was the longest tunnel in the world. Of the 1,250 navvies – navigational engineers or highly skilled labourers – who worked on the tunnel, twenty-six died – mainly in tunnel collapses and it was often plagued by flooding as Stephenson, worried about the effects of coal emissions on passengers, incorporated too many ventilation shafts.

One of the Stephensons' main rivals was Isambard Kingdom Brunel. Middle class and half-French, his seemingly genteel origins obscured a dramatic upbringing. His mother, Sophia Kingdom, was arrested as a British spy during the French Revolution and very narrowly escaped execution, while his father, the lesser-known inventor Sir Marc Isambard Brunel, was hurled into debtors' prison following several unprofitable projects. In *Lives of the Engineers,* Samuel Smiles describes the difference between the Stephensons and the Brunels: 'The Stephensons were inventive, practical and sagacious; the Brunels ingenious, imaginative and daring... Measured by practical and profitable results, the Stephensons were unquestionably the safer men to follow.'

Brunel came to fame early on in his career for his work on London's Thames Tunnel, a joint enterprise led by his father. It would become the first tunnel built beneath a navigable river.

Originally a foot tunnel connecting Wapping to Rotherhithe south of the river, the tunnel was adapted for trains in the 1860s and is now part of the East London line from Highbury & Islington to West Croydon on the London Overground. Work on the tunnel began in February 1825 when the younger Brunel was just eighteen and it opened to the public eighteen years later on 25 March 1843 – not an unusual time span for complex engineering projects.

While Robert Stephenson was working on the London to Birmingham line, the Great Western Railway Company's directors employed Brunel to build their new line from London to Bristol. The Great Western Railway (GWR) should, like Stephenson's London to Birmingham line, have terminated at Euston, however, because Brunel didn't want to use a standard gauge railway, opting for the wider seven-foot gauge instead, he built Paddington as its terminus. Had both lines ended at Euston, the station would have been far larger and Primrose Hill would likely have been flattened to accommodate more railway tracks instead of being developed into a pretty village – which might just have eased HS2's passage into Euston. Nevertheless, the GWR was another huge success. J. M. W. Turner captured the new technology's power in his 1844 painting, *Rain, Steam and Speed – The Great Western Railway*, which now hangs in the National Gallery. In the painting, a dark locomotive charges through a bank of fog; hiding in the mists is a terrified hare. Technology had outpaced nature.

By the 1840s, Victorian England was deep in the throes of railway mania. Previously, intercity transport consisted of horse-drawn carriages, in which you felt every bump of the road, while also being at the mercy of floods, reckless coachmen, exhausted horses and highway robbers. People were delighted by this new form of affordable public transport which drastically reduced journey times and made their journeys comfortable, safe and sociable. All rail companies had to do was ask parliament's permission to build, and since many MPs were investors, they were all too happy to oblige. While on the Continent, governments were more restive about this free-for-all, in Britain the government adopted a thoroughly laissez-faire approach. Not one prime minister suggested national planning and there were no checks on the economic viability of companies.

Quite a few rail companies competed directly, each running their own line on the same profitable routes. Companies, and company chairmen in particular, would spend vast sums on magnificent stations, symbols of their newfound influence. At the height of the boom in 1844, two railway companies, the Huddersfield and Manchester Railway and Canal Company and the Manchester and Leeds Railway Company, commissioned the magnificent neoclassical Huddersfield Station, designed by architect James Pigott Pritchett – a specialist in churches. Railways were the new religion.

Almost anyone with money was buying shares, from rich City lawyers and bankers to village vicars and chemists. There was a whole industry of railway magazines and newspapers to advise them. No one thought they could lose. As the mania took off, people traded increasingly on local and London stock exchanges, buying and selling shares in the opportunity to get rich quick. Lawyers and clerks who managed the stock-issuing made vast amounts of money, as did newspapers that took rail company advertising. Even engravers of share certificates prospered. In 1845, 240 bills were put to parliament representing 2,820 miles.

According to railway historian Christian Wolmar there is 'no equivalent in modern times of such a major investing scheme accounting for so much of a country's economic activity'. But perhaps you could compare it to one hundred per cent mortgages and the financial crash of 2008.

Then suddenly in the autumn of 1845, railway shares collapsed. A surefire investment turned into a liability as it became obvious that many railway companies were overspending on grand stations, exaggerating revenue and experiencing great difficulty building lines as quickly or cheaply as they had claimed. *The Times* and *Economist*, themselves products of the increased mobility of the railway age, were immediately blamed for the crash because they had run editorials highly critical of the behaviour of the railway companies. *The Times*, one of the most read newspapers in the country ran a leader in October 1845, the weekend before shares collapsed, observing that 'the mania for railway speculation has reached that height at which all follies, however absurd in themselves, cease to be ludicrous, and become, by reason of their universality, fit subjects for the politician to consider as well as the moralist.' A few weeks later, the paper also carried an 'exposé' about the speculators.

Cheap money – banking loans with low interest – collapsed the same year, as the British banking system fractured due to government attempts to control the money supply. The devastatingly belated decision to send vast amounts of aid to Ireland to alleviate the Great Famine, in which a million people died, also contributed to the end of credit which had been available for investment in railway shares, although Irish workers who had fled the famine continued to be a source of cheap labour for the rest of the century. Five years later, in 1850, railway investments were worth a third of what they had been three years earlier. Many in the new urban middle class lost everything.

In *A History of the English Railways*, published in 1851, John Francis describes the panic:

> It reached every hearth, it saddened every heart in the metropolis. Entire families were ruined. There was scarcely an important town in England but what beheld some wretched suicide. Daughters, delicately nurtured, went out to seek their bread; sons were recalled from academies; households were separated, homes were desecrated by the emissaries of the law.

Women investors constituted a surprisingly high proportion of railway stockholders. Even the Brontë sisters lost savings – including the first £500 Charlotte received for *Jane Eyre*. The company they invested in was owned by George Hudson, the 'railway king' – who raised vast amounts of capital for the York and North Midland Railway which opened in 1839. In 1844, he controlled a thousand miles of track – a quarter of the network – and was the elected Tory MP for Sunderland. A year later it all came crashing down. He had overextended wildly and kept expanding his business to pay off debts. When it came to light that he had been paying shareholders dividends from the capital he had raised (rather than the revenue from the railway), he could no longer continue to operate and fled to France to escape his creditors, losing his seat in Sunderland. When he was tempted back to England to fight for another seat in the 1865 general election, he was arrested by the sheriff of York and thrown into a debtors' prison for three months. The great commentator of the age, Thomas Carlyle, declared him little more than 'a big swollen gambler', while others admired his sheer chutzpah.

Nevertheless, the crash didn't put an end to the railways, which steamed on, albeit a little more slowly, for the rest of the nineteenth century. The crash did mean everyone was a lot less gung-ho about investing in unrealistic schemes because there was less money around. Large numbers of navvies continued to die and suffer injuries during the construction of the railways and after thirty-two men died and six hundred were injured during the building of the Woodhead Tunnel on the Transpennine Route from Sheffield to Manchester between 1839 and 1852, MPs set up a parliamentary inquiry. Navvies, the inquiry established, were forced to live in appalling conditions, first in wooden shanty towns and then in stone huts accommodating twenty of them at a time. There was no water nor means of sanitation and disease was rife. Despite the findings, Victorian England blamed the navvies' lifestyle of drinking and eating bad mutton for their suffering. Even as late as the 1880s and 1890s, some five hundred men a year were dying building the railways.

Building lines on the cheap was also dangerous for passengers. Despite a magnificent exterior, Huddersfield Station was notorious for its structural vulnerability and in 1885 the roof gave way, killing four people. But it was the 1879 collapse of the Tay Bridge in Scotland which demonstrated the perils of a penny-pinching approach. Caused by faulty engineering, it remains one of the deadliest rail disasters in British history. The bridge had been designed by Edinburgh engineer Sir Thomas Bouch, the Gerald Ratner of the late nineteenth-century railway industry, who boasted frequently and publicly about how cheap his construction costs were. He had miscalculated how the Tay Bridge would behave in high winds and shortly after it opened, the bridge broke in two during a violent storm. A crossing train plunged into the icy waters below, killing all seventy-five people on board.

There was some consolidation in this period as large railway companies took over smaller ones. From the 1860s through to the 1880s, the Great Western Railway purchased the Bristol and Exeter Railway, South Wales Railway, the West Midland Railway, the South Devon Railway and the Cornwall Railway. By the end of the century every small town in the UK had been joined up by a railway. You might have expected that after the 1845 share crash, the government would have intervened to exercise a sliver of state control. Instead, remarkably little regulation occurred; the laws that were passed mainly concerned fares and safety. Just before the bubble burst, the president of the board of trade, William Gladstone, introduced the 1844 Railway Regulation Act which controlled fares, defined what constituted a train and introduced compulsory daily trains for third-class passengers at a maximum cost of a penny a mile (known as 'parliamentary trains'). Railway companies were also obliged to cover third-class wagons, which until that point had been open-topped – as depicted by Turner – and to keep accounts.

More legislation followed as safety became a worry. In response to the Armagh train disaster which killed eighty people, including twenty Sunday school children, in what is now Northern Ireland, the government introduced the Regulation of Railways Act of 1889. The law made block signalling compulsory, regularised the interlocking of all points and signals and ordered all trains to be fitted with continuous automatic brakes.

***

Despite the railway share crash, Britain was relatively politically stable throughout the nineteenth century and none of the periodic wars, revolutions and instability which wracked parts of the European continent took place on British soil. As a result, European governments were more determined to exercise control over train routes, especially as the railways' strategic and military importance became evident as the century developed. Britain only needed to move large numbers of troops and munitions to ports to fight foreign wars.

The only context in which Britain did understand the military and strategic benefit of railways was in the colonies, particularly India, where railways were used to govern. The British built railways to carry cotton, iron ore and coal to Indian ports which were then shipped back to Britain. Later, they used trains to move troops quickly around the vast subcontinent to suppress rebellions. In 1853 Great Indian Peninsula Railway, a company financed by British MPs, bankers, cavalry officers and merchants, ran the first train from Bombay (Mumbai) to Tanna (Thane). Trains soon criss-crossed the country with tracks, locomotives and carriages for new lines manufactured in Britain and imported. Indians themselves had wanted to build and run their own railways and in the early 1840s industrialist and mine owner Prince Dwarkanath Tagore (grandfather of the Nobel laureate) raised the finance for a railway in Bengal. It never went ahead because the directors of the East India Company argued that rail was too important to be 'under native management'.

The Indian railways not only constituted a vast market for English train manufacturers: between 1854 and 1947, India imported around 14,400 locomotives from England, but India also provided British engineers, signalmen and mechanics and even railway employees like ticket collectors with massive amounts of work. Indians themselves were only allowed to occupy roles as navvies (where thousands died) or as menial repair workers. Following Indian independence in the 1940s, locomotive factories were established and Indians finally controlled their own network. Politician and historian Shashi Tharoor notes: 'The construction of the Indian Railways is often pointed to by apologists for empire as one of the ways in which British colonialism benefited the subcontinent, ignoring the obvious fact that many countries also built railways without having to go to the trouble and expense of being colonised to do so.'

Modern historians see railways as the 'tools of Empire' along with harbours, telegraphs and government buildings. Today, the Chinese, learning from the British, have even invested in the Uganda express, which joins Mombasa in Kenya with Lake Victoria. This railway was originally built by the British using Indian navvies to facilitate trade as well as to govern the area. US president Theodore Roosevelt, who travelled on the line while on safari in 1909, described his journey thus: 'The railroad, the embodiment of the eager, masterful, materialistic civilisation of today, was pushed through a region in which nature, both as regards wild man and wild beast, does not differ materially from what it was in Europe in the late Pleistocene [ice age].'

***

Meanwhile on the Continent, the two largest powers and landmasses were following the British and railway building accelerated from the mid-nineteenth century onwards. Railways were used, as in Britain, to power the growing industrial revolution and move people faster across the country. But railways also had huge strategic military importance, particularly after the rise of Prussia under Otto von Bismarck.

In France, the first proper railway line using locomotives was built between St Etienne and Lyon in 1832. A private railway company built the line to move coal into a growing industrial centre. The emperor of France gave permission, but that was the extent of state involvement. Once the success of the line was proved (it soon carried passengers as well as coal), railway companies started springing up around the country. Engineers used English technology and English iron to build the lines – that's why, with a few exceptions, French trains on twin tracks still run on the left-hand side. A small railway boom started; however, unlike in England, lines were primarily built to and from Paris rather than between other cities because industrial development in France was not nearly as advanced as in Britain. The first lines from Paris ran to St Germain (1837) and to Rouen and Orléans (1842). Commercial success was limited as most trains carried little freight and were primarily commuter lines to and from the capital; they were largely nationalistic projects to keep up with the old enemy in England.

Then, in 1842, French opposition politician Adolphe Thiers argued that the state needed to take control. A compromise was forged, balancing the advantages of the laissez-faire British system and the strict state control of the Belgian system. Train lines were franchised out to six railway companies. State engineers would plan and build the tracks and bridges, while these companies would manage the locomotives, rolling stock and timetabling. Some private companies still built railway lines in the provinces without going through government, but they became fewer and further between. The French engineers were educated by the state. Their railways enabled fast industrial growth from the mid-1840s onwards. Because the lines were run by the state, they were not nearly as vulnerable to financial shocks. As industrialisation accelerated in the second half of the century, the speed of trains captured the public imagination. Émile Zola's novel *La Bête Humaine* centres on the madness of a train engineer in the face of inhumane technological change. And yet the biggest weakness of the French system was a traditional one: most lines led to Paris, because the French believed Paris was not only the centre of France, but all Europe.

Meanwhile, Germany's railways were beginning to take shape. However, they were established very differently to those in France. Throughout most of the nineteenth century, Germany did not exist – in its place was a scattering of smaller states. Some railway lines were built by the private sector and others were run by the various duchies, particularly in the south. The Prussian railway system was private at first and then nationalised in the middle of the century, but the railways in Germany amounted to a network of lines criss-crossing the country. During the Franco-Prussian war in 1870, Prussia's prime minister Bismarck found the railways system gave the country a great strategic advantage: the east–west train lines meant troops and military hardware could be moved easily to France's border and then Prussian troops could seize control of the French railway lines to besiege and occupy Paris.

When Bismarck united Germany after the war, he implemented the lesson. Not only did having a government-controlled rail network unite the country and drive Germany's later industrial revolution – joining up the Ruhr mining districts and iron production to ports like Hamburg – but railways allowed Germany to wrest military and strategic dominance from France. For instance, the railway line built from Strasbourg to Wörth in 1876 joined the coalfields of Alsace–Lorraine (French before the Franco-Prussian war) with Germany, rubbing salt in the wounds of French defeat. A totally nationalised rail system was in place until the early twentieth century and helped Germany move troops in the First and Second World Wars.

But railways in Germany have an even darker association, which may explain the country's embrace of high-speed rail. The Nazi regime used the old trains to ferry millions of Jews, Roma, gay and disabled people to extermination camps in Poland. Until the advent of high-speed rail in 1991, long-distance cross-country railway routes in Germany and across Europe still carried echoes from that time. The Platform 17 Memorial, erected in 1998 in Berlin, was Deutsche Bahn's attempt to come to terms with the role played by German railways in the Holocaust.

Europe's different cultural history explains why politicians and governments have wanted greater control over railways than in Britain and have subsequently invested in training engineers and developing railway technology. Creating a truly European network, a long-time dream of the European Commission, has been trickier, with individual countries still keen to cling on to their national railways. There was some hope that the Ukraine war and the broader Russian threat might prompt European cities to look at their railway connections more strategically, but unstable governments, the Covid pandemic and the rise in the cost of borrowing made progress slow. In fact, Deutsche Bahn, in which the German state is still a majority shareholder, is now an international company running trains all over the world, including in Britain, and has overextended itself so much that domestic services no longer run on time. A source of national pride has become a vehicle of shame.

2

The Rise of London and the Motorcar

From 1845 to 1900, some £3 billion in today's money was spent on building British railways. By 1900, British railway companies employed 600,000 people – around five per cent of the population. The Grand Junction Railway company owned and administered the whole town of Crewe – the parks, houses and schools – until 1938. Crewe Alexandra FC's supporters are still called the Railwaymen. The trains themselves were run on military lines, the guards and inspectors patrolling the country in uniform, and held in some romantic regard as in novels like E. Nesbit's *The Railway Children*. By 1914 there were 23,440 route miles around the country and more than one billion passenger journeys a year, peaking at around two billion in 1920.

When the First World War broke out, the government took all the railways into public hands so they had more control of troop movements, manufacturing and the movement of armaments. Military leaders were astonished by the inefficiency of the nineteenth-century free-for-all. They discovered different train companies running services to the same locations and a fragmented freight industry. But the government was not keen to take on railways and nationalise them completely, which was happening in Europe at the time. Instead, railways were rationalised and the 120 railway companies were forced under four big companies which were given an effective monopoly on the areas of the country where they operated. Political leaders hoped this would make them more financially viable.

In hindsight, it was a missed opportunity to rethink the role of railways for the twentieth century as a public good. Politicians viewed railways in the same way as other declining heavy industries of the time and managed them accordingly: through amalgamation. The 1930s depression hit railways particularly hard. Fewer people travelled and less freight was being carried. According to Professor Simon Gunn's paper, 'The History of Transport Systems in the UK', the amount of coal carried on railways decreased by sixteen per cent between 1913 and 1937, while the companies operating in coal-producing regions saw traffic more than halve as production fell. There was a brief resurgence of railways during the Second World War, but by 1947, railways were no longer profitable enough and the new Labour government nationalised them. British Rail was born a year later. From this time onwards, the Victorian railway network was seen by successive governments as a financial burden – the railway industry as outdated as the coal and clothing industries they had nurtured. British manufacturing had moved on to cars.

Surprising as it may seem now, Britain's motor industry was far ahead of those in Europe and the government had built an unrivalled trunk-road system (though they were much slower building motorways). By the Second World War there were two million motor vehicles on the road. In 1951, private car ownership was relatively low at fourteen per cent, but there were large numbers of goods vehicles, mainly run by independent operators, undercutting the railways. Public bus transport also increased, and by 1932 one hundred local authorities ran bus services. There was plenty of nostalgia for past greatness, but little political will to find a modern role for passenger rail travel nor to exercise any central strategic control over how goods moved around the country. Professor Gunn calls this a 'predict and provide' transport policy which he argues prevailed from 1919 to the 1990s. 'Since its inauguration in 1919 the Ministry of Transport tended to be reactive rather than agenda-setting,' he argues. This meant there was no integrated policy for UK transport, with each mode of transport being considered on its own merits.

So railways were vulnerable to cuts, as they were expensive for the state to run and maintain. By 1956, railways were losing money every year and from 1960, the deficit was running at around £100 million a year (£2.5 billion in today's money). This was mostly because of the increasingly rapid decline of freight transport, which had made up half of railway profits. People might have still been using trains for trips to the seaside and shopping in neighbouring towns and villages, but this was not enough to make a financially viable business. It was at this point that Ernest Marples, transport secretary in Harold Macmillan's Conservative government, asked Dr Richard Beeching, chairman of British Rail, to prepare a report for the government to reduce losses and reinstate commercial viability.

Beeching was not asked to design a new railway system, rethink its purpose or consider different kinds of trains, but to reorganise and make cuts – which he did. The cuts to railway lines he proposed were the most savage any British government has made and Beeching earned the soubriquet 'Butcher of the Branch Line'. As a result, British Rail closed 4,500 miles of track and 2,500 stations. Towns and villages around rural England suddenly found themselves cut off from their neighbours. British Rail's excuse was that only a few people used most of the lines and they could take the bus instead. However, there were some glaring miscalculations. The Oxford to Cambridge 'Varsity line' via Milton Keynes was lost; direct lines between North and South Wales closed with trains forced onto circuitous routes via Shrewsbury; and the Great Central railway, which ran from Sheffield to London via Nottingham, Rugby and Leicester was axed due to being deemed unnecessary duplication. Over the last twenty years proposals including HS2 have sought to remedy the problems caused by these major Beeching closures with very limited success.

That being said, much of Beeching's report contained sensible suggestions about rationalising British Rail's freight arm, though that's not how the history books remember him. The cuts are etched into the English imagination as the beginning of the end of popular railway travel. The future poet laureate John Betjeman led the charge against Beeching's cuts, starring in BBC documentaries chronicling the last steam trains on branch lines. Although the old rogue didn't much like loud, polluting steam trains in his part of London and his heroines had moved into cars – Miss Joan Hunter Dunn finds love not on a train, but in a Hillman, parked in a Camberley car park surrounded by Rovers and Austins.

Some remember British Rail with fondness, but as the century went on, the government cut railways more and more aggressively, a Sisyphean attempt to make them profitable that never succeeded. Many services did not run on time, freight continued to move onto the road and most people chose to travel by car, to which they formed a large emotional attachment. National projects for grand railway schemes like a British equivalent to the French TGV to which the British might also have formed an emotional attachment viewed with suspicion and quashed. This 'make do and mend' approach to a Victorian railway structure meant that while engineering expertise was retained in repairing old-fashioned railways and upgrading signalling as well as very slowly introducing electrification, nothing new was being built. Most engineering talent transferred to the aviation industry or worked on schemes abroad.

The only place in the UK where rail building thrived from the 1990s onwards was London. While the industries and mines in the rest of the country were shutting down and with national infrastructure and engineering capacity declining, London launched itself as a global financial centre. The 1986 Big Bang deregulated the stock exchange and brought in electronic trading. The result was electrifying. London quickly became a rival to New York and the resources and energy of the country were pulled into the central vortex. The population of the city, which had been falling, started rising in the early 1990s. Men and women, rich from banking bonuses, clutched bottles of champagne and flooded the city's new restaurants, bars and clubs. Commercial developers realised that, with so much cash flowing in, they could make heaps of money in real estate.

With the Big Bang came another change: the regulation of banks passed from the Bank of England to the Securities and Investment Board (later to become the Financial Services Authority). The Bank of England had insisted that all banks be situated within ten minutes of the governor's office, but the new regulators didn't care, so a new financial centre could be built and the perfect place was the deserted Docklands area in the east of the city around Canary Wharf, which soon became the locus of a regeneration frenzy. Land there was plentiful and relatively cheap and the government was keen to help. The only thing lacking was transport. The Victorians hadn't lived there, so they hadn't bothered to plan any transport connections and boats and ships on the Thames had served for centuries to carry goods and workers. But if offices and banks were to spring up, besuited City yuppies would need trains and Tubes for their commute, and reliable ones to boot.

The first attempt at a transport connection into the Docklands was the Docklands Light Railway (DLR). The DLR, planned in 1984, was a joint venture between the then Greater London Council, London Docklands Development Corporation, government departments and London Transport. Ministers provided a £70 million subsidy. It was a modern but simple design – two thirds of the original line ran on pre-existing railway infrastructure, the stations were modular and the light trains simple and driverless. The network was a modest seven-and-a-half miles long and made up of two routes from Tower Gateway and Stratford to Island Gardens. The DLR was rapidly extended but it couldn't take huge numbers of passengers and seemed dinky to many developers, not up to their extravagant aspirations. As Canary Wharf skyscrapers shot up in the late 1980s, wealthy investors were keen for a bigger and faster Tube line, not only to the City but into the heart of government at Westminster.

In 1990, the Jubilee Line Extension (JLE) was approved. It would run from Westminster to Stratford. The then prime minister, Margaret Thatcher, had been reluctant to fund the JLE, but Olympia and York, the Docklands developers, promised her they would pay £400 million of the estimated £1 billion projected cost (in 1989 prices). From the beginning there were financing problems. Tempting businesses to the Docklands proved trickier than Olympia and York had imagined and in 1992 they went bust. By the time John Major became prime minister, the Treasury was cooling on the idea of the new Tube line. An engineer who worked on the JLE remembers how the Treasury tried to persuade John Major's government to scrap the extension. To salvage it, the contractors – keen for the venture to continue – summoned London's most famous architects and promised each of them they could design one of the eleven stations. They then asked them to slip into every fashionable party in London so they could rub shoulders with ministers and extol the virtues of the scheme. The *Independent*'s architectural correspondent, Jonathan Glancey, wrote a glowing article about the proposed gorgeous new stations and an exhibition was held to show off the designs.

The ruse was successful. An agreement with the developers was made and Major's government coughed up the cash. The stations, overseen by the award-winning architect Roland Paoletti, won scores of awards and they are still celebrated today, in particular Canary Wharf station, built by Norman Foster, and Westminster, designed by Hopkins Architects. Never was there a better example of gold-plating. If costs were really a factor, the government would have saved a lot of money by building eleven identical modular Tube stations.

Crossrail (now known as the Elizabeth line) was an even more ambitious project to join up the east and west of the capital and again was linked to London's growing wealth and sprawl, with the intention of bringing outer London areas closer to offices in the centre of the city. The project was given royal assent in 2007, just one year before the global financial crash, so was still part of an optimistic vision for London.

The seventy-three-mile line stops at forty-one stations and links Heathrow and Reading with central London in the west, ending at Shenfield and Abbey Wood in the east. Not all the track is new and Crossrail uses Great Western mainline track west of Paddington and Great Eastern infrastructure from Stratford. The construction was complex, including twenty-six miles of tunnelling under central London, twisting under schools, over Tube lines and under escalators, avoiding sewers, electric cables and gas pipes. Tunnels under the Barbican Centre and Soho's recording studios were designed with special floating slabs that floated the entire track structure on springs to minimise noise and vibration.

No financial lessons had been learned from the Jubilee Line Extension though and new stations like Whitechapel, Tottenham Court Road, Bond Street and Paddington were among the ten designed by different architectural firms. Unsurprisingly, the new stations are beautiful, but each came in vastly over budget. Whitechapel Station, priced at £110 million, cost the taxpayer £659 million. Crossrail was famously badly run – in the end it was delivered four years late and went billions over budget (from £14.8 billion to £19 billion). However, now that it's open, Crossrail is incredibly popular in the capital, increasing rail capacity, bringing commuting time down to forty-five minutes for one-and-a-half million people and fuelling regeneration all along the line. It helped that Crossrail was sponsored by Transport for London (TfL), which understood Tube and railway engineering, and that consecutive mayors of London – Ken Livingstone, Boris Johnson and Sadiq Khan – provided consistent political support.

While London now has what amounts to a rolling programme of fairly extravagant and beautiful urban rail infrastructure, the rest of the country was starved of rail funds. The exception, before HS2, was HS1, built only because government ministers realised (belatedly) that it looked awful to run a high-speed train at 300 kilometres (186 miles) per hour through France and the Channel Tunnel, then plonk it onto a regional railway in England where it not only ran at half the speed, but had to be slotted in between local stopping trains. It showed Britain up to the French as being country bumpkins, making ministers a laughing stock in Paris.

Thus, the Channel Tunnel Rail link (or HS1 as it was later called) was born via an act of parliament in 1996, designed to run between the Channel Tunnel and St Pancras station. On Michael Heseltine's insistence it curled east via Stratford to increase urban regeneration in east London. Mark Bostock, the chief engineer from Arup, recounts how it was conceived:

> We developed five basic principles. Firstly, the new infrastructure had to be built within existing motorway corridors wherever possible. Secondly, it needed to connect with the existing network to allow high speed commuter services to London to operate. Thirdly, it had to avoid Areas of Outstanding Natural Beauty (ANOB) and, if we could not, we should tunnel underneath them. Fourthly, we should identify suitable intermediate stations as a catalyst for regeneration and I am reminded that our alignment meant that London could select Stratford as a

winning Olympic Games venue. Finally, when the route entered the London tunnels, it had to be under an existing transport corridor if at all possible. There was no specific mention of speed as we knew it had to be designed over ground for 300 km per hour. That gave us scope to devise a railway which minimised environment impacts.

HS1, unlike HS2, was built by a private company, London & Continental Railways (LCR), which won the contract to finance and build the rail line. This was to be done through Union Railways, a British Rail-owned company which was transferred to LCR. LCR was supposed to raise £1 billion through an IPO (a stock launch) and another few billion through debt. It soon became obvious that this was not remotely possible. The Treasury were keen to scupper HS1 at this point and had it not been for John Prescott, who by 1997 had become Labour's deputy prime minister and secretary of state for the environment, transport and the regions, the line would have been scrapped. But Prescott was convinced of the idea of regeneration the railway might bring to Kent, so he agreed a scheme of government-guaranteed bonds to cover the cost.

The project was divided into two parts, the easy section from the Channel Tunnel to Fawkham with one large viaduct and one tunnel and the difficult bit into St Pancras. The idea was to learn lessons from the easy section. Everything went more or less to plan, but expense was added by lobbying from Kent County Council to drive HS1 through the centre of Ashford. The most difficult parts were renovating St Pancras station and driving eleven miles of tunnel under London. St Pancras was overseen by Ailie MacAdam who was brought back from the USA where she had been managing roads and bridges – 'huge interchanges, $300–$400 million of infrastructure.' Her project-managing skills and collaborative approach was vital to the project's success. HS1 opened on time in 2001 and to its £6.8 billion budget.

Despite all the difficulties, HS1 was a relatively simple rail project and was built as an extension of TGV Nord: a French-style railway with French signalling technology on parts of the line. 'It was designed to operate out of the box,' said HS1 Ltd's former chair Rob Holden, who was also briefly chief executive of Crossrail, 'we were not going to be the first anything.' Also, the line only went into one major city – London – and only passed through one county, Kent, where the Tory council leader Sandy Bruce-Lockhart was onside. One property was compulsorily purchased and the rest were bought by negotiation on a voluntary basis. Just two completely new intermediate stations were built, at Stratford and Ebbsfleet.

The first part of the line was opened in 2003 and the second part into London in 2007. Stratford International was never an international station (despite its name) and trains to Belgium and France no longer stop at Ashford International either. To recoup some of the costs, which the government had ended up paying, it sold the thirty-year line concession for £2 billion to the Ontario Municipal Employees Retirement System and the Ontario Teachers' Pension Plan. In 2017, the pension fund sold it on for £3 billion to a group of three infrastructure investors: HICL Infrastructure, Equitix and South Korea's National Pension Service. Each owns a third of the concession which runs until 2040. Both the Eurostar and Javelin trains to stations in Kent use the line, the Javelin providing a fast commuter service into London from Kent and the Medway towns. There is, though, a lot of spare capacity and plans are afoot to run trains direct to Germany.

HS1 was a discrete project. Although politicians and railway engineers had once dared to believe a high-speed line from the Channel Tunnel into London could be the first part of a wider network, those hopes were dashed when St Pancras was lighted upon as a terminus station. The reasons were mostly financial: there was no need to build a new station and British Rail could offload an under-used, ageing and expensive asset to maintain to London and Continental, while TfL was happy to link the Tube at King's Cross. Instead, all the government energy during the 1990s and 2000s went into how to make rail privatisation work. Thatcher had been sceptical, but the 1992 Conservative Party manifesto had promised that British Rail would be abolished. Trains would be run by a variety of private operators who would bid against each other to run services to different places. The track would be run by a separate private company. What could possibly go wrong?

Although rail travel increased from the late 1990s, the fragmentation proved disastrous in the long-term. The first cracks appeared early on. Railtrack, the private company which won the bid to run the tracks, sacked many of its engineers to increase profits, then failed to maintain railway lines or signalling properly. The consequences were deadly, resulting in the Hatfield rail crash in 2000 which killed four people and injured more than seventy. Unsurprisingly, Railtrack went bankrupt and the Labour government took the management in-house, creating Network Rail. The private operators did give the government a kick to improve some of the Victorian lines: Richard Branson who ran

Virgin trains for instance insisted on modernisation of the West Coast Main Line so he could introduce faster Pendolino trains to Scotland. But the Department for Transport didn't spend enough money improving the line and the partnership with Virgin was an unhappy one, with the service plagued by delays and cancellations.

Privatisation also fragmented services. Companies had to rebid for franchises every few years, often spending millions each time. The big lines could be profitable, although since 1997 three different main operators on the East Coast Main Line have handed back the keys – and the trains to Edinburgh and Aberdeen are now run by a consortium on behalf of the government. Meanwhile on smaller lines, particularly in the Midlands and the North, private companies barely broke even and had no incentive to invest and no power to insist on government improvements to the track. Running new trains on patched-up railway lines made some travel a bit more comfortable but also ran the risk of delays and expensive engineering work. British travellers soon became familiar with the term 'rail replacement services'. More and more people piled onto the existing, decaying network.

The Covid pandemic proved the final straw. Any profitability from train services vanished as passengers disappeared overnight and the British rail network has since been more or less under state control. The privatisation experiment failed; franchising was abandoned. The government contributed £11.9 billion to the day-to-day operations of British trains in 2023, just over half (fifty-two per cent) of their income and now, under the 2024 Labour government, trains are to be brought into public hands again.

The Treasury always hated railways, because rail travel accounts for only a tiny fraction of passenger miles, around nine per cent. When looked at in terms of individual trips, the number is even lower. Of 799 billion journeys in 2023/4, only 1.6 billion were on National Rail, mostly in the South East, and 1.2 billion on the London Underground and Glasgow Subway. The car, which accounts for ninety per cent of passenger miles, is the preferred mode of travel for most of the population, yet trains still seemed to eat up money as if there was no tomorrow.

Meanwhile, as Britain allowed its railways to fall into disrepair and redundancy, other countries were radically rethinking them for the twenty-first century.

3

A National Mission

Six a.m., 1 October 1964. Two high-speed bullet trains set off in opposite directions – one from Tokyo to Osaka, the other from Osaka to Tokyo. Just four hours later, crowds who'd gathered excitedly on the platforms welcomed the trains, perfectly on time. A journey that had once taken over seven hours was cut by almost half and within a year, another fifty minutes had been shaved off the route. Moreover, Ryotaro Azuma, then governor of Tokyo predicted at the opening ceremony that 'the opening of this Shinkansen linking centres of culture in the east and west of our country in four hours, will bring with it great benefits in all fields including economy, industry, culture and tourism'. The Shinkansen was to herald the second age of rail. Just nineteen years after the Second World War, Japan had pulled off an economic miracle and revolutionised mass transport.

Meanwhile, Britain was still manufacturing steam engines in Crewe.

***

The Tokaido Shinkansen – literally the 'East Coast route, new main line' – opened alongside the '64 Tokyo Olympic Games, deploying the era's most cutting-edge technology. The world's first high-speed railway, it was electric and ran on a segregated straight line without level crossings. There was no signalling by the side of the line – instead, the train would automatically brake if the driver went over the speed limit. The train would run on standard gauge instead of the ageing, slow and congested narrow gauge lines predominant in Japan at the time. Although most of the train's expense was met by the government, which also issued bonds, the World Bank lent Japan ten per cent of the cost and as a condition ordered Japan to finish the line and *reduce* the speed limit from 250 kilometres per hour to 200 kmph (which translates to bringing down the speed by around 30 miles per hour). Most importantly, the line was completed within five years. Despite problems with noise, the bumpiness of the line and relatively expensive tickets, eleven million enthusiastic passengers rode the Shinkansen train within the first three months; within the first three years, the number had reached one hundred million.

The route from Tokyo Station to Shin-Ōsaka via Kyoto was 515 kilometres (320 miles). However, Japanese National Railways soon started building extensions south-west, first to Okayama and Hiroshima, then on to Fukuoka on the southern island of Kyushu, which entailed boring a new tunnel under the Kanmon Straits. Meanwhile, improvements were made to the line including laying the track on concrete slabs over the viaducts and through tunnels. From then on, despite financial setbacks – Japanese National Railways faced insolvency and was privatised in 1987 – Japan built the equivalent of one new transit line every four years until 2004. With each line, trains and the lines became faster and more efficient. Japanese bullet trains are still the most reliable in the world, the average delay a mere forty-two seconds. They have also proved to be much more popular than short-haul flights and retain eighty-five per cent of the market share over airlines on the same routes. When bullet trains started whipping between Tokyo and the northern city of Aomori in 2010, Japanese Airlines (JAL) and All Nippon Airlines (ANA) were forced to reduce the frequency and size of their planes. The effect was even more catastrophic for airlines serving provincial cities along the Shinkansen, which cannibalised their routes completely.

One of the biggest attractions for Japanese politicians early on was that high-speed rail, as well as stimulating growth, made housing more affordable and eased congestion in big cities. Sixty years on, the Shinkansen is still a source of national pride. TikTok is full of selfies taken on trains, parents buy children Shinkansen-shaped bento boxes and there has not been a single fatality on the line. There have been many iterations of the high-speed line and it is still being built out, with plans to extend the Hokkaido line to Sapporo.

Japan dominated high-speed rail technology. It took almost twenty years for another country to follow Japan's lead and it was the French who took up the baton. In 1981, socialist president François Mitterrand decided that as part of his mission to bring France into modern times he would launch the TGV. This, he vowed, would ensure 'the fruits of progress would be shared by all'.

The TGV proved to have similar benefits for France as the Shinkansen had for Japan, although instead of running on segregated lines, the TGV used the old French rail network. That first train from Paris to Lyon ran on conventional tracks for the first one hundred miles, then on high-speed tracks before returning to conventional tracks on the approach to Lyon. Later, more high-speed tracks were added to speed up the journey even more. The hybrid model allowed the French to do much more mixing and matching than was possible in Japan (which had to introduce the mini-Shinkansen to run on their old tracks) and meant that TGV lines could be built incrementally with segments of high-speed line being added over the years to speed up journeys and ensure the whole country benefited more quickly from the TGV. They also used existing lines into densely populated urban areas like Paris and Lyon, so they didn't have to tunnel or knock down houses. Such a flexible network idea meant that when HS1 engineers designed their high-speed line through Kent from the Channel Tunnel, it made sense to go the TGV route. Other smaller countries around France like Belgium and the Netherlands run the TGV on their tracks too.

However, the biggest high-speed network in Europe does not belong to the French, nor even the Germans with their high-speed ICE trains, but the Spanish. Since the 1990s, Spain has built more high-speed rail tracks (roughly 2,500 miles) than anywhere else in the world outside China. Their engineers work so efficiently, they boast they can construct a high-speed line for just 17 million euros per kilometre (approximately £23 million per mile) – half the price of other European high-speed networks and ten times cheaper than HS2, which is costing £232 million per mile to build. Every part of the railway is now modular, including the stations, and the high-speed project has enjoyed consistent long-term support from successive governments. Ordinary Spaniards burst with pride today when they talk about their trains.

How did Spain achieve what seems to be utterly impossible in the UK? Well, first the history is different. Spain was a dictatorship under General Franco until 1975. During that time, the country stagnated, people were isolated from the world and other European countries and the train lines (as well as roads) became antiquated. It was as if Spain had been held in a 1930s time warp while the globe modernised culturally and technologically. Young people and the dashing young prime minister Felipe González were desperate to be part of the democratic world and to catch up. Expo '92, a world fair held in Seville celebrating five hundred years since Columbus sailed from Spain to the Americas, was the chance to do that. But people needed to be able to reach the fair – so, like Tokyo, which had built the Shinkansen for the Olympic Games – Spain laid high-speed tracks from Madrid to Seville.

At first, the Spanish used French TGV technology but soon they developed their own. The country now has a state-sponsored high-speed rail industry that specialises in building lines fast and cheaply. Bolstered by massive research and development grants, engineers from the Spanish railways are genuinely innovative and Spain has in turn nurtured a large private sector building and engineering industry to rival any other Western country. The EU has been vital to this project – high-speed rail in Spain has enjoyed more EU funding than all the other networks in Europe combined.

More than three hundred high-speed trains now traverse Spain every day and twenty-eight of the forty-two Spanish provinces are joined by a high-speed AVE line, accounting for two thirds of the population. And they arrive on time. A train from Madrid to Barcelona, roughly the same distance as from London to Edinburgh, takes only two-and-a-half hours. There are numerous cheap high-speed rail options including the AVLO, a no-frills high-speed train with no catering, offering tickets from as little as

€20 for the whole journey. Of course, there are many differences between Spain and the UK; notably, labour is cheaper. Much of Spain is also rural, so train lines don't have to pass through so many residential areas, making laying track a lot easier. But the Spanish face their own problems – mountain ranges cover large swathes of country, requiring more tunnelling and diversions which ultimately incur greater costs.

Ineco, the state-owned Spanish engineering company, acknowledges the country's advantages, but believes there are more strategic reasons for Spanish success. A recent report shows how a clear mission and long-term plan have been vital, simultaneously noting how much money was wasted steering HS2 through the UK's complex legal and governance system. The report also highlights how Spanish railways tended to have a standardised design, the cooperation of local government and a clear line of command, with managers staying on for the whole project and minimal changes to the project once approved. They also identified that making sure the pricing and frequency of trains were right was also important to ensure it made sense for enough people to use high-speed rail once it was built.

When coupled with better education, high-speed rail has increased the number of jobs and contributed to rising GDP and having a high-speed rail service has a positive effect on house prices – as long as the high-speed line joins up with other forms of local transport.

Would Spanish high-speed rail pass British cost–benefit analysis? What about the constant scrutiny from the Treasury and Britain's revolving door of government ministers? Probably not. Passenger numbers are not nearly high enough to justify the building expenses; some smaller stations have only one hundred people using them a day and older lines are not maintained as well. Planting a station in the middle of nowhere or in a far-off suburb doesn't help kickstart the economy. As for tourism, high-speed rail encourages more day trippers, but tourists don't hang about and spend the night, which affects takings in hotels and restaurants.

Wider societal issues can also sometimes affect rail lines. As the Spanish rail network was expanding in 2008 and a high-speed line was being constructed to join the cities of northern Spain, ETA (the Basque separatist movement) went green, demanding protection for the countryside, considered some of the most beautiful in Europe. Not only did the group attempt to sabotage the line; one of their number assassinated a businessman linked to the project. Today, more than fifteen years later, Spanish leaders are pushing ahead with a high-speed line joining the three major Basque cities – San Sebastián, Bilbao and Vitoria-Gasteiz – and in 2024 won a €430 million loan from the European Investment Bank to do so. The line avoids areas of environmental importance and is part of the Atlantic Corridor connecting Spanish, Portuguese and French railways. It will also bring the Basque region closer to the Spanish capital.

***

China has built the longest high-speed network in the world – 46,000 kilometres (29,000 miles) of track – over the last seventeen years. The speed of construction has been extraordinary, even by totalitarian standards. High-speed rail was a national mission too, joining up a large country which the government wanted to control, particularly in places like Tibet and Xinjiang. The network helped to join new mega-cities springing up around the country to turbo-charge the economy, allowing large numbers of people to move long distances between urban centres for work and leisure. High-speed rail also advanced China's global Belt and Road Initiative, developing infrastructure that could be sold to other countries to ensure dependence on China. China, a country with a deep understanding of the strategic importance of communication networks, is using and developing high-speed rail technology in similar, if slightly more subtle ways, to Britain in the nineteenth century. Britain controlled India and other colonies through train networks that could carry troops and goods around the country fast, while also retaining a monopoly on steam train construction, powering its domestic industrial revolution by forcing its dependents to buy British-made trains and all the track components.

China opened its first high-speed rail line, running from Beijing to Tianjin, seventy-five miles away, in 2008. Without the core technology to build their own trains, China paid companies Alstom, Siemens and Kawasaki Heavy Industries for their knowledge through technology transfer agreements. Trains and components were literally shipped to China with blueprints so they could learn how to build the trains in reverse. As a result, the first train and track in China was very expensive as engineers sought to perfect the technology. But the track was built quickly, in just three years, and soon expanded.

As in Spain, the whole design of the Chinese network has become modular. Every station in China looks the same, as do the viaducts. No attempt was made to build expensive 'iconic stations' across the country. Professor Linda Tjia Yin-nor from City University in Hong Kong, who has researched Chinese railways extensively, explains that the design of the trains and tracks

A NATIONAL MISSION

was pared down to the basics so construction would be as cheap as possible. China benefits from large open spaces and if the Chinese Communist Party decides to build infrastructure, there are very few institutional or environmental hurdles to overcome. People in the path of the train are evicted without compunction. The government has a policy of 'develop first and fix later', so people ask for compensation retrospectively. If a high-speed train track ploughs through a lake and deprives fishermen of their living, the government will find them another job eventually through a capacity-building programme.

The project has been largely successful and has overtaken Japan's and Spain's lines as the world's largest high-speed network. Untapped tourist destinations previously thought to be too remote have attracted hordes of day trippers. The government has also actively encouraged rail travel by drastically cutting short-haul flights. There may be military advantages to trains in China, too, attractive to a Chinese Communist Party increasingly keen on controlling its citizens. High-speed trains are also ideal for transporting large batches of military equipment and missiles without being detected.

Another more outlandish suggestion is that high-speed trains could be used as 'doomsday trains' not only to carry, but launch, strategic nuclear weapons. Yin Zihong, associate professor of civil engineering with Southwest Jiaotong University in Chengdu is the lead scientist on a national research project funded by the central government looking at high-speed rail's potential for military deployment. He and his team determined that a high-speed train built on reinforced concrete would absorb the shock of a ballistic nuclear missile launch better than a traditional heavy industrial railway.

The World Bank, which financed some of China's early high-speed rail, reported that its success was, as in Spain, down to long-term planning, consistent execution and coordination between provincial and local government. Nevertheless, even in a highly controlled country like China, there are places where high-speed rail doesn't work. At least twenty-four stations along lines are deserted because trains no longer stop there. In some cities, the economic benefits haven't materialised because the population has taken advantage of high-speed rail to up and leave. Having a high-speed line, as Professor Yin-nor points out, is 'not a guarantee everyone will win'.

Nor has the network been cheap. In 2021, the China State Railway Group publicly admitted to *Newsweek* that the company was in debt to the tune of $900 billion. High-speed rail, even in China with its huge population, doesn't make a lot of money. Once it's built, it requires maintenance and although fares in China have risen, they have had to stay within the reach of a sufficiently large proportion of the population to make lines worth building.

While China appears a success story, the regime has not found it quite as easy to translate its formula to other countries. A Chinese train project, the Whoosh, built in Indonesia between Jakarta and Bandung, suffered severe building delays in part because of Covid, but also due to disputes over land and financing. The line ran $1 billion over budget and opened four years late in October 2023. The Whoosh operates at a maximum speed of 330 kilometres per hour (or 205 mph; about the same speed as many Chinese high-speed trains) and has reduced the journey times between both cities from three hours to forty minutes, but look beyond the impressive figures and the Whoosh is more underwhelming. The train strands passengers on the outskirts of both cities and compared to the original plan, which was to traverse the island of Java from Jakarta to Surabaya, the Whoosh that emerged was severely truncated. For its troubles, Indonesia has acquired a short section of high-speed line and a large debt to the Chinese government.

***

Many of those who oppose HS2 fall back on the argument that high-speed rail is not appropriate for a small country like the UK. However, if that was the case, wouldn't countries with vast amounts of open land like Australia, Canada or the USA have adopted high-speed rail systems? They haven't. That they are all Anglo-Saxon regulated market-based economic models cannot be a coincidence. The so-called California bullet train scheme, which would have reduced travel times between San Francisco and Los Angeles from twelve hours to two hours and forty minutes with 220 mile-an-hour trains, has been as disastrous as HS2, with estimated costs rising to between $88 and $128 billion. A $9 billion bond from the private sector was raised in 2008 and a referendum held to greenlight the project, but the line is still far from being completed. Somewhat ironically, it's been held up in the courts over environmental challenges. As one commentator noted, environmental legislation meant to protect the countryside from fourteen-lane motorways was now being deployed against a much greener form of travel.

Mired in political wrangling, the construction and route has also been called into question. The *New York Times* reported that the line was 'never based on the easiest or most direct route'. Instead, the newspaper observed, 'the train's path out of Los Angeles was diverted across a second mountain range to the rapidly growing suburbs of the Mojave Desert – a route whose most salient advantage appeared to be that it ran through the district of a powerful Los Angeles county supervisor.' The first phase was scheduled to be completed in the 2030s (and there's doubt even about that timeframe) and run between two rural districts in the Central Valley, Merced and Bakersfield, described as the 'initial operating segment'. By March 2025, the cost of the starter line had risen from $22.9 billion to more than $30 billion and President Trump's transportation secretary, Sean P. Duffy, threatened to pull a $3.1 billion federal grant, blaming 'financial mismanagement' and throwing the future of Californian high-speed rail into doubt.

Australia has also failed to build a high-speed line even though it makes environmental sense. Every year, more than seven million people board an hour's flight between the country's two largest cities, Sydney and Melbourne. The train between the two takes eleven hours as it curves around corners totally unsuitable for a fast train. A high-speed line that could reduce train times and cut air travel would significantly reduce the country's carbon emissions, but construction would be hellishly complicated (which is why the line hasn't been built) – there are miles of unspoilt bush, the landscape is mountainous in parts and subject to complex land disputes. Currently, $500 million has been marked for a much shorter line from Sydney North to Newcastle in New South Wales which would reduce journey times from two-and-a-half hours to one hour. There are still doubts over its successful completion despite backing from the government. Philip Laird of the University of Wollongong has publicly wondered whether Australia, which has been discussing high-speed rail for forty years, holds the world record for high-speed rail studies with no construction.

If the Australians and USA can't build high-speed rail, surely the more communitarian Canadians have developed a proper network? The answer unfortunately is no. In fact, Canada has an abysmal record on high-speed rail. Despite numerous studies since Shinkansen was launched in Japan, not a single mile of high-speed track has been laid. Things may be about to change, however. In February 2025, the then prime minister Justin Trudeau announced a line from Toronto to Quebec City covering 1,000 kilometres (approximately 600 miles) called the Alto high-speed network. It will be Canada's biggest infrastructure project ever.

All these examples demonstrate that a high-speed rail network is expensive, needs vast public subsidisation and a governmental determination to be interventionist, and is unlikely to make a profit on all its lines. Trains need to be regular enough and fares low enough to attract large numbers of passengers. Governments with successful networks have developed them to fulfil geopolitical and economic goals. Japan is an example of how bullet trains have become popular national symbols of progress and advanced technology, wiping out parts of the short-haul aviation industry, connecting cities and islands to share economic wealth and making housing more affordable in its larger cities where prices were out of control. In Spain, high-speed rail has also brought disparate regions together, and with cheap tickets and faster travel times made people feel part of a connected nation. The Spanish, like the Japanese, have grown a global high-speed rail industry. You can understand why plans for British high-speed rail appealed to some politicians – in theory, HS2 offered unity, economic growth and the gloss of cutting-edge technology. It was also clear that it could just as easily go awry.

4

In the Public Interest

When Lord Andrew Adonis stood up in the House of Lords in August 2009 to announce HS2, he proclaimed the high-speed line to be 'manifestly in the public interest'. He boasted that 46 million domestic flights could be replaced by rail journeys and that HS2 would decarbonise British transport. It was time, Adonis told the assembled house, that Britain caught up with the rest of the world on high-speed rail and stopped this attitude of 'make do and mend'.

A high-speed line to the North had been debated for years. When the French TGV first got off the ground in 1981, Britain had been developing similar technology, but instead of going all out for a TGV-like train which would need new train tracks, British Rail developed a fast diesel train – the InterCity 125 – which could operate on the existing tracks and relatively straight East Coast Line to Edinburgh. To increase speeds on the West Coast, where there were a lot more bends in the track, British Rail invested in an APT (advanced passenger train). The APT was electric, had the ability to tilt around corners and so increase speeds up to 160 miles per hour.

Tested for a decade from the mid-seventies, the APT cost £45 million to develop but as rail historians have chronicled, British Rail was not set up to complete a project which was so technologically innovative and there was neither the political will nor the management capability. Engineers were marginalised, working in out-of-the-way sheds in Derby. Later on, there was negative press – early versions were nicknamed Queasy Riders because of the travel sickness the tilting caused. Although a more stable APT-P ran for a year in 1985, Margaret Thatcher cancelled production and British Rail switched its focus to InterCity 125s. The endless testing had produced prototypes which other countries exploited – for instance, the Italian-made Pendolino introduced by Virgin trains on the West Coast Main Line fifteen years later uses the same APT technology, but more efficiently applied.

The effective transport strategy in the UK at the time (though mostly unspoken) was to maximise use of the Victorian rail system and spend new transport money on building roads and motorways. In such a small country, many powerful voices argued, there was no room for upgrading both.

However, Britain being Britain, neither was done properly and as the mid-noughties approached it was obvious the transport network outside London and the South East would collapse. Even the roads were overcrowded. In the early 2000s, there had been rumblings about high-speed rail. John Birt, the ex-head of the BBC brought in by Tony Blair's government to 'think the unthinkable', had suggested a TGV link from London to Scotland and the government then tentatively committed to looking at it in the 2005 manifesto, though Blair and his transport ministers at the time were lukewarm about embarking on another rail project without backing from the private sector.

To fulfil the commitment, the Department for Transport and the Treasury employed Sir Rod Eddington, a plain-speaking Australian. Eddington was a friend of Rupert Murdoch's who had made his career in the aviation industry – at Cathay Pacific and then as CEO of British Airways, where he had been responsible for mothballing Concorde. A man able to take tough decisions, he had little expertise in railways or railway-building. The remit was 'to advise on the long-term impact of transport decisions on the UK's productivity, stability and growth'. He was the perfect person to lay to rest any aspirations for high-speed rail. And with the guiding hand of the Treasury and Department for Transport (who are said to have actually written his report in 2006) he didn't disappoint.

> The risk is that transport policy can become the pursuit of icons. Almost invariably such projects – 'grands projets' – develop real momentum, driven by strong lobbying. The momentum can make such projects difficult – and unpopular – to stop, even when the benefit:cost equation does not stack up, or the environmental and landscape impacts are unacceptable.

Eddington said that the current infrastructure needed to be used more efficiently instead, a policy derided later by Lord Adonis. Like Birt, he identified a capacity issue. For example, there was not enough room on existing transport links – rail or road – to deliver economic growth and link up cities and ports for freight and passenger journeys. His solution: road pricing, charging people for using roads and motorways so they would use them less. Birt had suggested this too, in combination with a high-speed line. Eddington later clarified at a 2007 transport committee session that he wasn't against high-speed rail in principle for the 'densest corridors' but that 'high-speed rail with unproved technology and with dubious economic benefits is not something we should be spending £30–40 billion on'.

As with Birt, the road-pricing idea went down like a lead balloon. Even now, the idea of road pricing can cause a political attack of the vapours. It is extensively used in London through the congestion charge and ultra low emission zone (ULEZ), preventing polluting vehicles coming into the capital. Similar measures like low traffic neighbourhoods restrict driving and impose fines for contraventions. Other cities have introduced road pricing too, including Birmingham, Bath and Oxford. It's not always popular. Labour's 2023 by-election loss in Uxbridge was blamed on the extension of the ULEZ and the mayor, Sadiq Khan, faced pressure from the national Labour Party to scrap it. He didn't. Making individual drivers pay has helped fund the expansion of public transport.

On national roads, there is little appetite for road pricing. There are tolls on a small section of the M6 in the West Midlands and payment is required to use the Dartford Crossing. A landmark case in Scotland over charges to use the new bridge to the Isle of Skye led to the abolition of all pricing for crossing bridges and going through tunnels north of the English border.

Sir Rod's report also recommended a transport strategy for the long, medium and short term – as one might have in a business – so the Department for Transport would have an idea about what it was spending money on rather than reacting to problems as they came up. Although a thirty-year transport plan was subsequently written, it barely mentioned high-speed rail. Ruth Kelly, the Labour transport secretary – the third in as many years – welcomed the report and nothing happened.

And so it seemed that high-speed rail was dead in the water. However, the idea would be resurrected by the Tories under their new leader: David Cameron.

***

In 2008, the Conservatives had decided environmental policies would win them the next election. It made a toxic and 'nasty' political party look like the cuddly future. While some Conservative thinkers were running around talking about creative destruction, Cameron was filmed hugging a husky in the Arctic, unequivocal proof that he cared about melting icecaps. The Conservative logo was even changed from a torch to a tree. Against this background, high-speed rail suddenly became attractive. It was nominally 'green' and could replace domestic air travel. This would mean no third runway at Heathrow, which was agitating Cameron's west London base. The shadow transport secretary, Theresa Villiers, announced HS2 at the Conservative Party conference in 2008. Her argument, loosely based on a report by the engineering firm Atkins from March that year which had said a high-speed link would cost £31 billion, estimated the cost at £20 billion and that it would cut 66,430 domestic flights a year. Her train would go from St Pancras station to Leeds, Birmingham and Manchester. Villiers told the *Guardian*:

> This is a seriously green decision. A few years ago, it would have been inconceivable for the leader of the Conservative party to say no to a third runway and putting the brakes on Heathrow expansion.

There was uproar in the aviation industry, which was quick to condemn the new line because of the ditching of the third runway. So-called business leaders were not keen either and Scotland was angry that it wouldn't extend beyond Leeds. Ruth Kelly, who had just agreed a third runway for Heathrow, condemned the proposal, saying it was 'politically opportunistic, economically illiterate and hugely damaging to Britain's national interests.' This was not, she argued, a straight choice between air travel and railway capacity. Both were needed. She accused the Conservatives of being 'lightweight, shallow and only interested in grabbing a headline.' The Treasury came up with costs which estimated the Tory plans would cost £80 billion, not the £20 billion described – from the very beginning, estimated costs were being randomly plucked out of the air by civil servants to make political arguments.

High-speed rail also had support from the Liberal Democrats under Nick Clegg. Among the radical policies in their 2008 transport plan were a high-speed link to Heathrow from St Pancras and a line to Birmingham and Leeds, all to be built in separate stages through injections of private funding and taxes on domestic flights. The plans also included cutting fuel and vehicle excise duty and replacing those taxes with road pricing on motorways and major roads to encourage people to use public transport.

By this time, there was a ballooning political consensus that high-speed rail was essential, but that road pricing was too politically difficult and the aviation industry needed to be protected. Cake-ism (the idea that you can have everything you want without having to make any painful choices) was baked in from the start. No surprise then that HS2 was quickly characterised by opponents not as green or encouraging traffic off the roads, but as a white elephant, a macho way of keeping up with the continentals in a country which was becoming increasingly sceptical of the Europeans and their fancy bureaucratic doings. This view was encouraged by the car industry and their advisers. The Royal Automobile Club (RAC) brought out a report saying it would be better to improve the roads and the existing railway tracks than build something new.

But of course, Labour still held a significant majority at the time, so all these proposals were merely that.

***

Time for railway buff Andrew Adonis to make his entrance, stage left. Tony Blair's former policy adviser had recently been appointed to the House of Lords and decided now was the time to push HS2, staking his career on it.

By this time, Gordon Brown had taken over from Tony Blair. After some initial relief at his appointment, his premiership started to go wrong. It began when Brown pulled back from calling a general election after becoming prime minister in 2007. His public vacillations weakened him considerably. Then a year later, the first financial crisis hit with the shocking collapse of Northern Rock. The building-society-turned-bank had aggressively expanded and taken on subprime American mortgages which now couldn't be paid off. It was the first time a run on the bank had happened since the nineteenth century. Lines of people were filmed queuing outside branches of Northern Rock, desperate to retrieve their money.

In the end the government bailed out the bank and guaranteed people's savings, but worse was yet to come. The bankruptcy of American investment bank Lehman Brothers a year later left global finance in chaos, wiping billions off the world's balance sheets. Money which banks claimed to have was wrapped up in complicated investment vehicles and didn't actually exist – it was effectively a pile of worthless IOUs. For the UK this was catastrophic: the previous eleven years of the Labour government had been based on building up banks and the City of London and for Brown, who had been chancellor of the exchequer, it was personal. He had taken tax revenues from the City to fund government spending – but failed to regulate. City institutions were not only vulnerable to financial shocks, but they were *making them happen*. Brown reacted by deflecting – making himself the centre of global attempts to retrieve the failed banking system – without holding the City to account.

All of this brought home just how much the country's economy relied on financial services in the South East. The government hadn't exactly ignored the north of England or Scotland: a lot of regeneration of city centres had happened and shopping centres had been built from Glasgow to Birmingham. Brown had used taxes raised from the City to fund tax credits and increase public services. Jobs had been created by moving public bodies out of London, whether that was basing a tax office in Newcastle or moving the BBC to Salford. However, what Brown and the Blair government had neglected to do was to build up new industrial economies in the Midlands and the North, not only in towns, but in major cities.

Adonis's opportunity came when Ruth Kelly's career was suddenly terminated at the Labour Party conference, shortly before Villiers' high-speed rail announcement. She had fallen out with Gordon Brown over abortion – she was a strict Catholic and suspected of plotting against him. Her departure was announced in the bar of the Brighton Grand by Brown's macho press adviser Damian McBride (now a special adviser at the Home Office) at 3.15 a.m. to a band of mostly inebriated journalists. Brown confirmed it the next day. The post of transport secretary was suddenly vacant. Geoff Hoon (later to be revealed as another anti-Brown plotter) was quickly appointed to the role, but more significant was the appointment of Lord Adonis who, by his own account, had asked to be given the job of rail minister. Brown's advisers were surprised the prime minister had kept the Blairite Adonis, but Adonis had transferred allegiance.

Adonis was obsessed with railways. As a schoolboy he had campaigned for more rail services and written to the chairman of British Rail suggesting 483 timetable alterations. When he became rail minister, his first act was to travel up and down the country on trains writing a simultaneous blog for *The Times*. He believed in opening more local railway lines closed in the sixties and seventies, restoring stations and improving services. But most of all he wanted to introduce high-speed rail to Britain. He had travelled on high-speed trains around the world and was passionate about it.

He set up a fully funded separate company called HS2 with Geoff Hoon in January 2009. The records for this company are still in Companies House, as are those of similar companies established at the same time by civil servants from the Department for Transport: HS3, HS4, HS5 and HS6. HS3 was the next part of the high-speed network proposed by Adonis in 2016, another high-speed line across the North from Liverpool to Manchester, Leeds and Hull and ending up going north to Newcastle. The others remain a mystery.

The Department for Transport and the Treasury were sceptical about Adonis's plans, but they had allowed their ministers to go out and sell a railway line they expected to kill after the imminent general election. It was exactly the kind of madcap scheme that politicians came up with towards the end of their time in government, the sort the next government could be persuaded to drop. The then permanent secretary at the Treasury, Nicholas Macpherson, now Lord Macpherson, told me: 'There's nothing to lose at that point, and you are kicking the can down the road in terms of spending.'

Likewise, efficiently funding and overseeing innovative, high-cost projects wasn't the DfT's forte. Metronet, a public–private project designed to revamp and maintain the London Tube, had recently gone into administration. The department had also just been saddled with repaying and servicing HS1's project debt of £4.8 billion which they had guaranteed and which the company that had built the railway (London & Continental Railways) had found they couldn't pay off. To cap it all, they had also approved London's Crossrail (now the Elizabeth line), which was coming in at £2 billion over budget before so much as a sod had been turned.

Although HS2 was supposed to be a separate company independent of the DfT, at the beginning it was staffed by people seconded from the department. The chairman was the ex-permanent secretary Sir David Rowlands who had led on aviation policy at the DfT. He had got himself in a pickle by accepting a senior directorship at the British Airports Authority, the owners of Gatwick and Heathrow, straight out of the civil service. The government had blocked his appointment and he was offered HS2 as a consolation prize. In an interview with the *Evening Standard* he joked that he had been in the DfT for so long that 'if you want to blame someone for the state of the transport system, blame me.' The chief executive role was not advertised publicly. Instead, it was 'given' to Alison Munro (her own words), who had begun in the civil service as an economist.

Hoon told the House of Commons he had asked the HS2 company to:

> Report by the end of the year with a proposed route from London to the West Midlands, setting out any necessary options, including for stations. It will also consider the potential for new lines to serve the north of England and Scotland.

His announcement received support both from the Conservatives and the Liberal Democrats. He was asked to expand the line further north by MPs but gave little further detail.

***

The man who was put in charge of designing HS2 was former Network Rail engineer Andrew McNaughton. A jovial, bearded man with huge experience on British railways and now a professor at Southampton University, he quickly got stuck in. The brief was to design a standalone ultra-modern, high-speed line with its own digital signalling which would last at least sixty years and travel at a maximum of 400 kilometres per hour (approximately 250 mph), a speed far higher than HS1 and high-speed trains in Europe (which usually travelled at around 300 kmph or 186 mph). The trains would leave from central London and reach Birmingham, Manchester, Sheffield, the East Midlands and Leeds with a possible spur to Heathrow airport and a connection to the HS1. Some would also be able to travel on conventional lines so passengers could reach other parts of the country.

The line was to cause minimum disruption to existing transport links and destroy as few houses as possible. This was to minimise expense and so the DfT did not have to relive the trauma of the West Coast Main Line (WCML) upgrades when passengers had spent almost a decade in a variety of rail replacement buses with little tangible improvement. The old WCML would take regional trains, stopping commuter trains and freight.

McNaughton was to determine the route and work out the kind of trains which would be needed as well as come up with the budget. He started work in February 2009. He explains what happened:

> Day one, I had no phone, no computer, we had a rickety desk in a bit of surplus government property at 55 Victoria Street where the SRA [Strategic Rail Authority] used to live!
>
> I was seconded from Network Rail. I was the chief engineer there and so this was a melding of rail knowledge, which was me, with transport planning and economic knowledge, which was Alison [Munro]. Various people were seconded from the DfT and I also sought out a small number of people that I trusted completely to come and support me. Then we went out and got, by competitive tender, Arup to be our engineering consultant and Temple to do the environmental stuff.

The deadline was by the end of the year. So there we were, on the 31 December, trying to get the printers to work at five in the evening because our secretary of state (now Lord Andrew Adonis) had made it very plain that he was going to spend New Year's Day reading it. The report was about 200 pages long and it had got in it principles that have been with us ever since, such as this is a project to maximise economic benefit.

So it's demand-led – not engineering-led, not operational-led – it's demand-led. The reason you run trains off to Manchester on day one by going up to the West Midlands and then using the conventional existing West Coast Main Line is because of that first principle, which was to maximise the benefit.

***

Before we go any further, it is worth understanding why Gordon Brown backed high-speed rail. While he was chancellor, Brown had been very opposed to high-speed rail, considering it expensive and a waste of public money. According to Patrick Diamond, then director of policy planning, it was the financial crisis, subsequent recession and the need to look beyond financial services that changed Brown's mind. The effect of the banking crisis was stark. Brown writes in his autobiography that tax receipts for 2009/10 were projected to be £616 billion. Instead, the Treasury received £452 billion.

The government could no longer rely on the City of London to bring in revenues to keep the country going and debt was rising. Brown explains that he wanted to adopt a Keynesian strategy of pumping more money into the economy to stimulate growth, but that the Treasury and the Bank of England thwarted him, believing that balancing the books was more important. But there were measures that Brown was able to take in what Diamond describes as 'post-crisis' policies which would move the country away from financial services and into manufacturing. Similar initiatives throughout the following decade were to be called the 'Northern Powerhouse' (George Osborne) and 'Levelling Up' (Boris Johnson). Brown's label was the more technocratic phrase: 'rebalancing the economy'.

There were several initiatives Brown wanted, including a car scrappage scheme and 'catapults' with universities. These were not the medieval sling mechanisms, but rather the idea that growth happens when universities are encouraged to 'catapult' their research into business and make money off it. Cambridge had been very successful in doing this through both high-tech and life-science organisations like the mobile phone chipmaker Arm and AstraZeneca. But catapults don't excite the public and car scrappage only gives a short-term boost to the car industry. High-speed rail, on the other hand, is tangible, strategic and looks like the future. The fast rail link to the Channel Tunnel had been opened in 2007 and the Javelin domestic service in 2009. St Pancras station was an enormous success – the perfect blend of old restored railway sheds, shops and new platforms. All this was going through Number 10 strategists' minds as they contemplated how to bring the idea of 'rebalancing the economy' to life.

Adonis, who replaced Hoon in mid-2009 as secretary of state, was more than happy to have full-throttle prime ministerial backing for his HS2 project. He was fanatical about HS2, with a chart in his office showing route options out of London. Within two weeks of getting the job, Adonis went to Number 10 and presented a paper to Brown. His proposals were more coherent than anything Hoon had described. As one of Brown's advisers, Gavin Kelly, remembers: 'You could see in Gordon's eyes how excited he was about what Andrew was talking about.'

This was not only about rebalancing the economy, it was also politically important in the short term to present plans to connect London with Birmingham because of the seats in play in 2009 in the Midlands. Those Labour constituencies were looking increasingly marginal, as polling showed voters deserting the ruling party. Brown was right to be worried, but wrong that his HS2 announcement would help: Labour would go on to lose fourteen seats in the West Midlands at the subsequent 2010 election. Ministers were onside too – when Adonis presented his high-speed rail proposals to the cabinet he received a standing ovation.

Unlike Brown, Adonis was just as interested in getting Tory buy-in. As he tells it, he nobbled Cameron about HS2 at Charlbury Station, a Cotswolds village outside Oxford. 'I opened the new platform as transport secretary, with the local MP, David Cameron, in 2009,' Adonis recalled. 'I went with him afterwards to his constituency cottage and showed him the plans for HS2 and urged him to make it a cross-party project.'

Adonis may well be misremembering – the meeting likely took place when Adonis was rail minister and visited Charlbury in March 2009 as the first spade was turned for the platform project. If that's true, Adonis was consulting with Cameron before he had officially received the cabinet's support.

***

If kickstarting the economy in the North was the political motivation behind HS2, then it is astonishing that so few questions were asked at the time about how exactly high-speed rail might do that. Of course, stations might bring housing and more jobs, and HS2 itself would create a major pipeline of engineering projects which would upskill the workforce and employ British businesses for years. McNaughton allowed himself to dream that it might even allow Britain to become competitive in the global rail engineering market again.

However, most manufacturing in the North had disappeared and the productivity gap with London was huge – around thirty-five per cent (which it still is). Most of Britain's old industrial base had been run down during Margaret Thatcher's and John Major's tenures, and the rest left to quietly collapse under New Labour. A fifteen-year project for a high-speed train was not going to rectify that kind of decline unless there was an integrated national industrial strategy to build up the economy of the North, local people with the skills to drive it, enthusiasm and funding from private-sector investors and the close cooperation and buy-in of regional politicians, who might even have been encouraged to put some money behind HS2. But no such strategy was laid out, because HS2 was seen as a Department for Transport project and, as its name suggests, the department concerned itself only with transport.

And if the aim was to help the North, why was the train line even going to be built from London? Richard Leese, leader of Manchester City Council at the time, argued the case, and today, Henri Murison, the CEO of the Northern Powerhouse Partnership – a large group of northern businesses – agrees with Leese, but the North in 2009 held little sway in Westminster and had no strong independent voice. In Victorian times, train lines *were* built between northern cities and ports before being expanded down to the capital, because the North with its coal mines and cotton mills was the financial engine of the country. By 2009, it was London driving the economy and so it was logical in politicians' minds that any major rail project should start there.

But there was a further consideration: HS2 had to be submitted to a Treasury cost–benefit analysis. London is far richer than the rest of the country and that makes the figures look better. Even joining Britain's second, third and fourth wealthiest cities appears bad value for money because the land prices outside London are relatively inexpensive and with the economy so much weaker, the returns would also appear low. It's a ridiculous way of calculating the kind of projects a country should invest in, and in practice has meant the UK only builds serious infrastructure in London and the South East, creating a devastating feedback loop – but the rules of the Treasury's so-called 'green book' still dominate decision-making on infrastructure.

***

Brown was enthusiastic and the opposition was onside so, in March 2010, Adonis came to the House of Lords with a command paper. He announced HS2, a line costing £30 billion that would run 140 miles from Euston to a new station in Birmingham and then, splitting into a Y shape, trains would continue north-east to Leeds and north-west to Manchester, before joining up to the existing East Coast and West Coast mainlines to Scotland. A total of 335 miles of line would be built. Trains from London would reach Birmingham in thirty and fifty minutes, depending on the stations used, with Manchester, Leeds and Sheffield all brought to within seventy-five minutes of London. Journey times between Birmingham and Leeds and Birmingham and Manchester would be halved to forty-five minutes. From the moment Adonis stood up and presented McNaughton's plan, the broad outline of the route was set.

The argument Adonis made to the Lords was light on 'rebalancing the economy', and the boost to the northern regions was almost a footnote. Instead, Adonis spoke about capacity on the line, speed between London and the North and integrating the Victorian network. From the very beginning, there wasn't much clarity about what the new railway was actually for and that meant the objectives shifted over time. Not having clear goals, according to Bent Flyvbjerg, author of *How Big Things Get Done*, is fatal to any project – from kitchen conversions to complex rail projects.

Instead of challenging Adonis to debate the need for high-speed rail, the noble Lords gave him an easy ride. Only the Tory Lady Hanham, who was supportive of high-speed rail, expressed scepticism that the plans were fully developed, saying the statement was 'full of general principles but lacks detail on cost analysis, funding sources and the expertise necessary to deliver any part of the project'. The Liberal Democrat peer Lord Bradshaw was more fulsome, suggesting that Adonis had already 'got his foot on the lowest step' of the 'pantheon of railway greats' like Brunel and Stephenson.

No one at this stage seemed very worried about the costs and meanwhile at the Treasury, plans to scupper HS2 after the election were already afoot.

But there were billions of pounds at stake. HS2 didn't have a clear rationale everyone believed in, plans to go beyond Birmingham were sketchy at best, speed was over-emphasised and HS2 wasn't linked to any kind of wider industrial strategy, even though rebalancing the economy had been the main political reason for going ahead with it. Not only that, but Britain was going it alone. While Adonis had been inspired by high-speed networks abroad, and McNaughton had spoken to engineers in Europe, there was no team from Spain or France regularly advising the government on the lessons they had learned.

And then the details of the route kept being modified. HS2 was always part of a broader struggle to work out how to level up Britain which continues to this day. The political and economic foundations upon which this high-speed line was being planned were already shaky.

5

Rebellion in the Shires

When the newly elected prime minister David Cameron announced in 2010 that, instead of dumping HS2, the coalition government was going to plough ahead with it, a howl of despair rang out through the Tory heartlands. The new transport secretary Philip Hammond, a rather lugubrious patrician figure, unused to meeting the public, was promptly dispatched to win over despairing communities along the line to Birmingham. The same people who had just returned Conservative MPs appeared in force in village halls, greeting the minister with anti-HS2 placards. A rather shell-shocked Hammond was filmed by the Channel 4 news crew he had injudiciously brought along, declaring that he could move the line a few hundred metres here and there, but he couldn't re-route it completely.

While it hadn't mattered to Gordon Brown that HS2 ran through sixteen Conservative constituencies, it was trickier for Cameron, especially as he hadn't even won the election outright. He didn't have solid cabinet support either – five of his ministers represented affected areas. But he was an optimist and a commitment was included in the coalition agreement – forged after the election between the Tories and the Liberal Democrats – to develop a high-speed network.

The main driver of HS2 was Cameron's chancellor George Osborne, and HS2 wouldn't have happened but for his enthusiastic support at every stage. He shared Brown's vision: HS2 could be a route to jobs, growth and connecting cities. In 2013 he enthusiastically told the BBC about the project: 'Time and again, we have this debate in our country about how we're going to bring the gap between north and south together, about how we're going to make sure that our growth is not just based on the City of London. High-Speed 2 is about changing the economic geography of this country, making sure the North and the Midlands benefit from the recovery as well.' For Osborne, HS2 proved an antidote to the severe austerity he was imposing on the country as part of his vow to wipe out the £109-billion budget deficit, racked up by the previous government to bail out the banks.

Philip Hammond soon grew fed up with HS2 and the number of people who kept buttonholing him to re-route the line to protect their country piles. The project was eating swathes of Hammond's time and could prove career-ending should it run aground. Although HS2 was supposed to be a green project, environmental concerns were central to the opposition. Every well-respected institution in the country from the National Trust to the Campaign to Protect Rural England had registered objections and an array of powerful landowners were creeping out of the woodwork. The railway commentariat, led by author and blogger Christian Wolmar, had also lined up against HS2.

So Hammond decided the best thing to do was to announce a £50 million compensation scheme for members of the public who were going to be affected and kick HS2 firmly into the long grass. Although the strategy for HS2 from the beginning has essentially been 'decide, announce, defend', Hammond decreed that he was going to launch a lengthy consultation – the largest in British history, as his successor Justine Greening later boasted – to discover what the public really thought. For a year and a half, HS2 staff and Department for Transport officials trekked up and down the proposed length of the line from London to Birmingham, holding numerous public meetings and gathering responses. Around 55,000 individuals and businesses replied from 'across the country' – you can still find the consultation report online. Researchers found that most people were against HS2, concerned about the cost, environmental impact and value for money. Perhaps unsurprisingly, some of HS2's greatest supporters were unionists in Scotland, who believed they could benefit from faster train links to London, along with those in the South East, and pro-Europeans who liked the idea of connecting northern cities with Paris and Brussels. Many commended the link to Heathrow Airport.

Meanwhile, to the surprise of civil servants, Labour council leaders in northern cities were quite indifferent, dispirited and distrustful, worried more about impending austerity cuts to their budgets than a fast train link to the capital. Legislation for metro mayors for the West Midlands and Greater Manchester had not yet even been introduced into parliament.

***

The stretch of countryside most affected by the HS2 route to Birmingham was the Chiltern Hills, a designated area of outstanding natural beauty (AONB) in Buckinghamshire. The HS2 consultants chose this route precisely because it was unspoiled; the government would save compensation money because they wouldn't have to knock down many houses and businesses. Other possible routes, following the route of the M1 for instance, had been discarded precisely because they were so built up. The report acknowledges some 'landscape impact', but digging up large swathes of the home counties was judged a price worth paying.

Though the engineers had envisaged some tunnelling, they had decided that miles of the line could run through open countryside, ploughing its way through picturesque villages, across golf courses and over cricket grounds. Most of it was to run straight through the constituency of Chesham and Amersham. When local Conservative MP Cheryl Gillan first found out in 2009, she was incandescent. Her constituents quickly piled into her Friday surgeries to voice their dismay. Why, they asked Gillan, did restrictive AONB planning rules prevent them building garages or extensions, but the government could turn up and construct a bloody great railway?

Action groups sprung up in every village along the route in Buckinghamshire. Large signs were driven into the ground outside cosy British pubs, anti-HS2 banners stretched across hedgerows and farmers' fields. At first, the outrage had been anti-Labour, but the anger was now turned towards the new coalition government.

Emma Crane, previously a corporate lawyer, had moved her family from London to an idyllic little house in South Heath, a small village near the proposed line. She couldn't believe politicians had chosen her area – she and her husband, a lawyer, had bought a house there to enjoy English rural life. They had a large garden and envisaged their children growing up in the peaceful tranquillity and safety of a beautiful village. It was the perfect spot – her husband could commute to London and continue his high-flying career in the city.

'Why destroy that part of Buckinghamshire? You couldn't have chosen a more beautiful place if you tried,' she told me.

Meanwhile, the Stop HS2 action groups argued their communities were going to be torn apart for a railway which would simply become a commuter line between London and Birmingham. During Christmas 2010, local families turned up at Chequers, the prime minister's country house, with a blow-up white elephant.

Members of the local Conservative Association actively coordinated rebel meetings, which Gillan encouraged. Her counterparts in neighbouring constituencies were keeping their heads down. Some had already managed to squeeze concessions out of Hammond. Fellow Conservative MP Andrea Leadsom had saved the village of Brackley in Northamptonshire from demolition and Jeremy Wright, a government whip representing Kenilworth, had been gifted a covered tunnel. David Lidington, the centrist MP for Aylesbury who, like Gillan, had a constituency at the heart of the route, had decided it was safest not to raise his head above the parapet in case he was forced to resign as minister for Europe.

Cheryl Gillan, then in her fifties, demonstrated less caution. The daughter of an ex-army major and a WREN who had served in the Second World War, she was determined to both rebel and retain her new job as secretary of state for Wales. Gillan, elected in 1992, was one of only twenty female Conservative MPs in parliament at that time. To succeed in that environment and bag a safe seat in Buckinghamshire, you had to possess the hide of a rhinoceros and a will of cast iron. Gillan had both. In her 2021 obituary, it was noted that early on in her career when she was sitting in London traffic, a mugger had opened the door of her Range Rover Discovery and grabbed her handbag, containing her House of Commons pass, from the passenger seat. Gillan had leapt out, heels abandoned, setting off in pursuit in her stockinged feet.

One of her contemporaries described her as a 'forceful woman'. He said she could be quite acute about people and slightly overbearing and other Conservative MPs, including Cameron, were terrified of her – which might explain why he didn't sack her sooner. When she was a whip, she insisted on parliamentary protocol. One rather cerebral MP found himself rebuked as if he were her dog for not nodding at the right moment in parliament. 'I'm trying to train you,' she informed him in an accent honed and tautened at Cheltenham Ladies' College.

While she didn't quite have what it took to become a top political player, the former City marketing executive with her blonde power bob proved a highly effective campaigner, capable of making important cross-party allies and politically organising in her constituency, not something male Tory MPs traditionally felt was their role. The railway, she argued, was going to be environmentally damaging while bringing the locals no advantage because it didn't stop in any towns along the route.

That's not to say there would have been more support *had* HS2 stopped in Buckinghamshire. One civil servant recalls offering the leader of Buckinghamshire County Council a train station in

Aylesbury, but was told in no uncertain terms that a HS2 station would mean more housing and that would be hugely unpopular.

Gillan realised that, as the route was never going to be changed, it was essential to ensure the line went *under* the Chilterns rather than *over* them. Every mile of tunnelling added hundreds of millions of pounds to the cost of the HS2, but that didn't concern Gillan or her constituents. And as the years went on, she saw no contradiction in working with the chair of the Public Accounts Committee (PAC), Margaret Hodge, to interrogate rapidly rising costs.

When the line was first planned, only around six miles of track were due to be underground – a stretch between the M25 and Old Amersham, before following the A413 road. Gillan told the *Sunday Times* in June 2011 that 'if the project goes ahead I would resign the whip unless the prime minister tells me he would allow me to vote against it.' Cameron, she continued, had known of her opposition when he appointed her as secretary of state for Wales. Gillan was sailing close to the wind, even closer than many imagined, when it emerged that she had employed a lawyer, Mark Vivis, specifically to campaign against HS2.

As the date of the HS2 announcement approached, Cameron laid the ground for Gillan's resignation, briefing the newspapers that he had lined up another MP, Maria Miller, to take over from her. Gillan had upped the ante, infuriating Cameron, by encouraging the chairman of her local Conservative Association to write to all Tory MPs encouraging them not to support HS2 and to 'work with us to see this project off.'

In January 2012, the big announcement came. Justine Greening released a written statement to the House of Commons that HS2 would go ahead. Her reasoning included economic growth, joining up cities and capacity on the rail network, though not specifically on the West Coast Main Line.

She promised the line to Birmingham would be completed by 2026 as part of Phase 1. Phase 2 would comprise of high-speed links between Leeds and Manchester with 'intermediate stations' in the East Midlands and South Yorkshire. There would be direct links to Heathrow Airport and to the Continent via HS1 (the Channel Tunnel link).

It would all form the basis of a larger high-speed network with lines allowing direct trips to Wigan, Preston, Liverpool, Newcastle, Glasgow and Edinburgh. On HS2, journey times from London to Edinburgh and Glasgow would be cut by over an hour to only three and a half hours. The whole package, Greening announced, would cost only £32.7 billion in 2011 prices, generate £47 billion in economic benefits and fare revenues of up to £34 billion over a sixty-year period.

An accompanying press release rammed home the argument, stating: 'there are no credible alternatives to a new railway line'. The figures were wildly optimistic. As Channel 4's fact checker noted, the Greening announcement used different figures than those in the report which accompanied it – £34.6 billion – and failed to take into account that the overall cost of building *and* operating HS2 was even higher at £58.1 billion (important for accurate cost–benefit analysis). All the figures were calculated at 2011 prices and in the next few years would be revised upwards again and again.

For Gillan and other concerned MPs, the secretary of state produced sweeteners in the form of large amounts of extra tunnelling – a fifty per cent increase on the existing plans. There was to be more deep tunnelling and so-called green tunnels elsewhere (tunnels built in trenches across fields then covered over). An extra 1.5 miles was promised under the Chilterns, taking the line underground to Little Missenden, which gave Gillan an excuse not to resign. According even to HS2's optimistic take at the time, the cost of the Chilterns tunnel was to be around £171.4 million (based on the estimation that every seven kilometres – or 4.35 miles – of tunnelling cost £500 million), though in truth, underground conditions are so unpredictable that no one can accurately assess how much a deep tunnel will cost through modelling alone.

Greening had been persuaded by campaigners that the noise of the train for local villages would be similar to the noise of planes flying over her west London constituency into Heathrow. McNaughton claims today that she insisted not only on tunnels but on high, solid sound barriers along the tracks in mitigation. So encased was the HS2 line set to be that there would only be views for seven per cent of the entire London to Birmingham journey. When civil servants remonstrated, they were told it would be possible to install moving pictures on the HS2 trains' windows to simulate the countryside.

Greening tried to persuade the House of Commons that the extra tunnels would actually save hundreds of millions of pounds; an assertion met with derision from Labour. Gillan, to the prime minister's dismay, had found another reason not to resign, but Cameron used a reshuffle a few months later to sack her. According to contemporaries, she didn't leave quietly, shouting at Cameron, who hated conflict, when he told her he was replacing her as Welsh secretary with a man. To discredit her completely, journalists were alerted to the fact that she had sold her house near Amersham, set within 500 yards of the proposed route, just months before Greening's announcement.

Even if it was the end of her ministerial career, Gillan was now free to become a full-time thorn in the government's side. She consistently stood up in the House of Commons to oppose HS2 and briefed against the high-speed railway to whoever would listen. For her troubles she managed to squeeze out of the government another two and a half miles of tunnel from Little Missenden to South Heath – and underneath the M25. Her constituents loved her, but her punishment was increasing isolation on the Tory benches and the opprobrium of civil servants, who identified her as public enemy number one.

***

Politicians were taking it upon themselves to change the route and the scope randomly by a few metres while in town hall meetings with local residents, or with the addition of a tunnel here or there when faced with political opposition.

There were other problems. How could the government determine an accurate price if engineers hadn't worked out how the train could link up with the Eurostar? Nor were there detailed plans for a spur to Heathrow, which was a fantasy even then. Little thought had been put into the line beyond Birmingham to Manchester, now relegated to a Phase 2 development, let alone how trains would run on existing lines to Newcastle and Edinburgh. There was no timescale for the train to run beyond the West Midlands.

Engagement with the public had barely begun. A hybrid bill had yet to be written, let alone circulate through parliament. Only when such a bill was passed could contracts be procured for building the line and designing the new trains.

6

Clearing a Path Through the City

While Cheryl Gillan was succeeding in extracting miles of expensive tunnel for the wealthy inhabitants of Buckinghamshire, civil servants were working out how to drive HS2 straight through housing estates in Camden to reach a new HS2 terminus in Euston. And while some of the poorest people in the borough would lose their homes under HS2's plans, many thousands of others nearby were destined, if the train line went ahead, to spend decades on building sites, the vibrations of diggers, the roar of lorries and the deep thud of piles being sunk the new soundtrack to their lives. For locals, the prospect of reaching Birmingham twenty minutes quicker in what they perceived to be a luxury train designed primarily for businessmen wasn't an enticing prospect. Nor were they particularly interested in freight capacity problems in the North. They just wanted to keep their homes.

Nasrine Djemai remembers exactly when she found out in 2009. The room in the community centre was packed with anxious neighbours from the three grey blocks which make up part of Camden's Regent's Park Estate. The leader of the council,

Sarah Hayward, the local Labour MP Frank Dobson and some suits from the Department for Transport were there to talk to residents. 'There were older residents, people in their eighties and nineties who had fought for their country and were suddenly told their houses were going to be demolished.' They kept asking: 'Where are you going to move us?'

After the meeting, she went back to her flat with her father. He was dismayed. Having escaped the dictatorial regime in Algeria and set up home with his family in London, he was suddenly facing another random authoritarian act from Britain. The plans were hypothetical, unimaginable and the family were afraid but determined to fight back. The rumour mill was in full swing across the estate: there was talk of being moved out and with the huge housing shortage locally, 'Where do you plan on putting three-hundred-plus dwellings? It didn't seem even possible,' Djemai, now an elegant 28-year-old, reflects.

***

Euston was never an ideal place to put a new HS2 station or bring in completely new high-speed railway lines. There was an unsightly and crowded station there already which was too small to be adapted and the surrounding land was densely populated, home to more than 10,000 people. Although there were some elegant Georgian houses, most people lived in council flats, some in 1950s high rises on the Regent's Park Estate, built to replace housing destroyed by the Blitz, and others in early twentieth-century social housing in Somers Town. A Labour council for most of its existence, Camden had, as best it could, resisted pricing these residents out of the centre of the city. Nevertheless, the area hadn't completely escaped gentrification and the nearby King's Cross development represented gentrification on steroids.

Euston had several other constraints. The station was flanked by some important institutions – the British Library and the soon-to-be built Francis Crick Institute – while one of London's major arteries, the Euston Road, runs south of the station. So what possessed the engineers and civil servants? Like all unsatisfactory solutions, Euston seemed like the least bad option. This was to prove a gross miscalculation.

***

Andrew McNaughton, the chief engineer, had considered at least twenty-five alternatives, ranging from the relatively sensible to the absurd. The most outrageous proposal was to dig up St James's Park along with part of Buckingham Palace, or to use the estate as the landmark station. Unsurprisingly, this idea was promptly scrapped, the DfT keen to avoid outright treason. Another suggestion was to build a terminus under Regent's Park which would join up to Baker Street Tube station, offering passengers convenient access to the Bakerloo, Jubilee, Circle, Hammersmith & City and Metropolitan lines. However, the operation would have entailed digging up the entire park and then turfing it over. Moreover, set in one of London's wealthiest neighbourhoods, the sheer number of rich people and judges living around the park likely to launch legal action was enough to deter even the DfT.

Then, there was Hyde Park: if there are underground car parks, why not a train station, the engineers reasoned? But TfL didn't like the idea, because there was nowhere for the 60,000 likely train passengers to continue their onward journey. Even a station under the Thames by Westminster was floated but eventually struck off due to expense and disruption. Paddington, Broad Street and Bow were all ruled out because the land around them had been sold off for offices and commercial developments. Marylebone was considered too small.

Had some strategic thinking taken place a couple of decades earlier, King's Cross, 750 metres away, might have been an option. In the late 1980s Arup had looked at deserted goods yards and locomotive depots to the north of King's Cross as a potential HS1 Channel Tunnel Rail Link station, positing that trains might be able to travel on from Brussels up to Manchester via the West Coast Main Line. But by the time HS2 planners rocked up, all the old derelict railway land and buildings had been sold off for the King's Cross development, one of the largest and most fashionable new neighbourhoods in London. There wasn't space to run any train lines there, or expand the station.

***

At first, senior Camden council officers didn't take the DfT officials who knocked on their door in 2009 very seriously because the plan seemed so ridiculous. There were always random people turning up at the council with grandiose ideas. But very soon they realised McNaughton and his men meant business and were proposing to drive eighteen high-speed trains an hour right into the heart of the borough and were somehow going to carve out space to build a mega-station.

***

Djemai was right to be concerned about the fate of her family and their neighbours. Since the railways first came to central London in the nineteenth century there has been an unfortunate record of demolishing poor people's homes to bring in the tracks and expand the station. Charles Dickens in *Dombey and Son* describes the Euston area at that time:

> [It was as if] a great earthquake had, just at that period, rent the whole neighbourhood to its centre. Traces of its course were visible on every side. Houses were knocked down; streets broken through and stopped; deep pits and trenches dug in the ground; enormous heaps of earth and clay thrown up; buildings that were undermined and shaking, propped by great beams of wood.

A century later in 1938, the London Midland & Scottish Railway pushed out hundreds of people to expand the station again; many ended up in Dartmouth Park in north London, housed in what are still referred to as the railway flats.

By 2010, Euston was a little different in that democratically elected local politicians were not going to allow their residents to be moved out of town. As in the shires, anti-HS2 groups sprang up all over the borough. Rather than pursue a collaborative approach with the council (as Ailie MacAdam had done with HS1), those running the HS2 project very soon began to rile people with their rhetoric and what was perceived as a high-handed attitude. They didn't believe in consultation – instead they held confrontational meetings where they told people what they were going to do. Angered by this approach, Camden Council leader Hayward was determined from the beginning to pursue a strategy of non-cooperation and extract as many concessions from the government and HS2 as possible to ensure that locals could stay in the area.

She had massive support from her residents, and Camden's sometimes unruly tenants (who were not always as enamoured of the council) were delighted by the fight. Hayward's chief executive, more used to working with people across London and finding a compromise, was rather taken aback by her antagonistic attitude to government officials and HS2 bosses, but as one of her other top officials, David Joyce, remembers, her methods were certainly effective in the short term.

The Department for Transport, under intense pressure from Camden Council and their MP – first, Dobson and later, future prime minister (and future Sir) Keir Starmer – eventually funded new blocks. When they were ready, her family, Djemai told me, were given one week to vacate their old home, chased by HS2 contractors who had started blocking doorways with large plastic orange barriers and erecting scaffolding. A couple of weeks later, their high rise, Ainsdale, along with two others, Eskdale and Silverdale, housing three hundred people, was demolished.

While HS2 had reluctantly agreed to pay for rehousing council tenants and leaseholders whose blocks needed to be flattened, the company was not about to pay money to or buy up the homes of the thousands of other Londoners who lived in the area and were likely to be bringing up their families on a noisy, polluting building site for the next couple of decades – not much different from Dickens' time. In rural areas, by contrast, HS2 had either compensated or bought the houses of people who lived up to 120 metres away from the tracks just because of fears about noise and vibrations. The generous offer was the result of intense lobbying and a judicial review and meant that HS2's compensation bill was rising well beyond what had been budgeted for. Village dwellers, golf club owners and farmers along the route were effectively draining HS2 dry. The total compensation bill for the first phase of HS2 between London and the West Midlands was to come in at £3.6 billion. The idea that council tenants in the city centre might be afforded the same was not on the table even though the building works were to occur on their front doorsteps. Instead, there was an assumption that thousands of city dwellers, particularly the least well off, were used to noise, dust and pollution. Likewise, it is clear some Conservative politicians felt that the poor had no place in central London and shouldn't be subsidised to live there.

Drummond Street, an important hub for the local South Asian community, lined with Indian and Bangladeshi restaurants, fell victim to HS2's uncompromising approach. The restaurateurs mounted a huge campaign to stop one end of their street being boarded off, preventing travellers from Euston station coming to eat there. In the end they received no compensation, although £650,000 was made available to renovate the street. A Catholic girls' secondary school, serving some of the most disadvantaged children in the borough, was also demolished and the Bree Louise Irish pub was flattened by the railway, destroying some of the last vestiges of Irish culture in the centre of Camden.

A few wealthy and influential people with the ear of government did manage to persuade HS2 Ltd to buy their houses – notably Stanley Johnson, father of the London mayor and future prime minister Boris Johnson. He kicked up a fuss, claiming that his house, which was relatively far away, would be affected by train vibrations and parking and access problems because of the construction works. He also claimed he qualified under the 'need to sell' scheme because he wanted to move closer to an elderly relative. After several years of lobbying, the government bought his Regency house in the leafy enclave of Park Village West for £4.4 million in 2016. Only three other local homeowners managed to pull off the same trick. Most were less lucky. Nick Cam from Park Village East told his local paper, the *Camden New Journal*, that HS2 had refused to buy his house despite him putting in three applications.

***

The treatment of the dead – rich and poor – was the most careful, as they were categorised as heritage assets. From early on it was clear that the St James' Gardens Burial Ground and park, adjacent to Euston station, would be needed for construction works. The old graveyard contained 60,000 bodies, buried there between 1790 and 1853. In 2017, HS2 employed hundreds of archaeologists and conservationists to embark on the biggest exhumation in British history. The most noteworthy find was the body of Captain Matthew Flinders, who died in 1814 and led the first western European circumnavigation of Australia. The Wellcome Collection believed the site gave a unique insight into the lives of all classes of people at that time. The Camden site was the most spectacular of all the historic sites along the route, but between 2018 and 2020, HS2 employed more than a thousand archaeologists, specialists, scientists and conservators up and down the line to Birmingham, the sector's biggest employer at the time.

Although HS2's liaison team and senior executives spent huge amounts of time dealing with the living, attending often acrimonious community forums and liaison groups, locals always had the impression that these works were being imposed from on high. HS2 and the DfT only conceded ground if they were forced. It was easy for the population and all the other bodies with an interest in the area and transport (from TfL to Network Rail) to be obstructive. Indeed some, like Frank Dobson, certainly felt that making life as difficult as possible for HS2 might persuade the government to terminate the train at Old Oak Common instead, as he certainly didn't see any benefits for his constituency. Had the bosses at HS2 pursued a more collaborative approach, there might well have been a much better outcome for everyone, but they didn't. Euston was set up from the beginning to be a mess.

Djemai still remembers the feeling that overnight her family and their neighbours had become 'just guests in the area we have lived in our whole lives'.

> I very much understood, I came from a certain socioeconomic background that wasn't a level playing field. It made me essentially understand that anything could be done to me because I was the child of council tenants. There wasn't very much agency in what could and couldn't be done to us. We tried to broker that agency when we said, 'we are not going anywhere'.

7

Grand Projet Angst

'We knew it was bollocks from day one,' Margaret Hodge, former MP and chair of the Public Accounts Committee from 2010 to 2015, tells me over the phone. I can make out the chatter of her grandchildren playing in the background. Now a Labour peer, Hodge is no-nonsense and sceptical of large expensive schemes. She sniffed a grandiose overblown project out quickly with HS2.

The Public Accounts Committee is one of the most powerful committees in the House of Commons. It was set up by William Gladstone in 1862 to scrutinise government spending. Professor Peter Hennessy, Britain's top constitutional expert, described it as 'the queen of the select committees… [which] by its very existence exerted a cleansing effect in all government departments'. The committee has the power to examine the finances of any government department or project and has a wide-ranging remit. The chair must be from the opposition. Until 2010, members of the committee were appointed by their respective political parties, mirroring the proportion of seats they held in parliament, and then those members elected the chair. It meant chairs tended to be Establishment figures, approved of by whips who didn't want to cause too much trouble either to the government or to the leader of the opposition. In the 148 years before Hodge took over the committee, every chair had been male.

In 2010 this all changed with the so-called 'Wright reforms'. Designed to open up parliament and give backbenchers more power, the reforms meant committee chairs were elected by all MPs. For Hodge, it was a great opportunity. There were several men, she says, who believed the role was 'theirs', but it was she who put in the time persuading MPs. One third of whom, she remembers, were new to parliament. She spent ten days canvassing in the House of Commons tearoom, the neo-Gothic café deep in the House of Commons reserved for members of parliament.

Before Hodge came on the scene, the PAC was not a political committee and was under the strict discipline of the comptroller and auditor general – the head of the National Audit Office (NAO), Amyas Morse. He didn't like that Hodge brought a dramatic political edge to the committee. Before her disruptive influence, he ran a tight ship where he decided what questions to ask and what projects to examine.

'It was totally controlled by the NAO,' said Hodge. 'We were just the end of the sausage machine.' She decided to break with tradition and plough her own furrow. She sought independent advice on how to hold the government and civil service to account for their spending. HS2 was in her sights and she was assisted by her three wise men: Tony Travers, a local government expert, Steve Bundred, formerly head of the Audit Commission, and David Walker, the *Guardian* journalist. They would meet secretly before any committee to decide the best target for scrutiny.

Many of the questions the committee asked of HS2 were documented in three reports during this period: What lessons had been learned from previous projects? What was the evidence that HS2 would lead to regeneration on the route? How robust were the costs of building it? How had the costs and benefits of HS2 been calculated? Did Britain have the engineering skills to carry it out? How was this part of a strategic rail plan?

Hodge maintains to this day that civil servants lied to justify HS2, although she stops short of identifying anyone. The permanent secretary at the Department for Transport, Philip Rutnam, she assured me, was more honest than many, ready to put up his hand when he was wrong and less prone to obfuscation than others. But, she says ruefully, 'It was a *grand projet*... and there are men who like *grands projets*.'

Capacity, one of the main reasons for the line, was not as much of an issue as it was made out to be, Hodge argues – although this view is much disputed. Committee members believed that modelling overestimated the number of passengers who would use the line and rubbished the idea that travel time on trains is dead time (thus, shortening journey time is good for the economy) because people work on trains. Rather than see the engineering industry's backing and the promise of tens of thousands of jobs as a positive, there was suspicion from many Tories that the line was being built simply to line engineers' pockets. As Conservative MP Richard Bacon said during a committee hearing in 2013: 'If the project goes ahead, the industry gets to employ lots of people and the government pays it lots of money. The industry has an interest in this project going ahead.'

Hodge's main problem was that it was difficult to work out what exactly was going on and who she could trust. In 2010, the engineering firm Atkins had, the committee ascertained, 'double-counted' the benefits of HS2 by £8 billion, while conveniently neglecting to factor in the cost of disruption to businesses in central London and rebuilding Euston. By this time the civil service was marshalling all its expertise to push the bill through parliament – but the PAC was still rehearsing whether HS2 should be built at all.

In their 2013 report, the committee writes:

> The Department has yet to demonstrate that this is the best way to spend £50 billion [*sic*] on rail investment in these constrained times; that this is the most effective and economic way of responding to future demand patterns, that the figures predicting future demand are robust and credible and that the improved connectivity between London and regional cities will enhance growth and activity in the regions rather than drawing more activity into London.

There's a video of the committee hearing in 2012, ostensibly about lessons learned from HS1. The transport officials are being grilled about why they had only come up with an evaluation plan for HS1, retrofitting the case for it years after it had been completed and why they had miscalculated the passenger numbers so badly. Couldn't they have foreseen there might be competition from cheap airlines and the Channel ferries? Then, Hodge pivots to HS2:

> It looks to me as if your High-Speed 2 rail calculations change annually as well. In 2010, you expected demand to double by 2033; in 2011, you expected demand to double by 2043; and in

2012, we are back and doubling by 2038... all of this leaves us with very little confidence in the Department's capability, built on the record of what it did on HS1, of predicting properly the crucial factor of passenger demand when you take such a massive investment decision.

To which there is no real answer. Hodge then moves on to the actual money. At this time, HS2 was still only set to cost a mere £30 billion, though everyone suspected the estimated costs were rising. Then permanent secretary of the DfT, Philip Rutnam, bespectacled and earnest, tries donnishly to explain that the plans for HS2 are 'right up there with best practice in dealing with the costs of major projects.' Hodge, with one hand on her chin and the other on her hip, looks sceptical, as do the committee men arrayed around her, posed like a slightly bedraggled magistrates' bench. The civil servants sit directly in front of them as if in a courtroom dock.

Hodge starts: 'The Major Projects Board or whatever they call themselves in the Cabinet Office, have they vetted your...?'

Another eager civil servant jumps in: 'Yes, we have been through the gateway process with the Major Projects Authority.' It's the answer Hodge has been looking for. 'And what have you got, red, amber, green?' The official is forced to answer: 'Amber/Red', before launching into an explanation of how they hope to change this when they are letting out the contracts. To which Hodge declares with fantastic hauteur: 'Amber/Red is not good enough... Amber/Red is not a state we aspire to stay in: it's a state we aspire to get ourselves out of.' The civil servant quivers.

Hodge says that after the performance at that first committee in 2012, she was approached by Cheryl Gillan, who would feed her information for the hearings that were to come. 'We worked on it together,' she said. They liked each other. Hodge describes Gillan as 'Tory, Tory, Tory' and 'a doer'. Both were defying their party. Although Labour wobbled about whether to support HS2 between 2013 and 2014, the then Labour leader told her he wasn't terribly happy with her grandstanding. 'I did get into a bit of trouble with Ed Miliband. For criticising it,' she tells me.

The most serious accusation Hodge levels is that the civil service was unfit to commission and oversee a project of the scale and complexity of HS2. 'Lack of capacity in the civil service to manage major projects', she told me, 'was one of the most persistent issues that we were forced to focus on.'

'Traditionally the civil service was a thinktank and was about dreaming up policy, and it was those people who were rewarded within the service and promoted. It wasn't about implementation – that was considered to be a second class,' she notes.

It was HS2 Ltd, the wholly government-funded company, that oversaw implementation, with the Department for Transport 'the client'. However, in reality, both were inextricably entangled and operated to the political demands of successive ministers. With a lack of engineers among its staff – they employ only a handful – the DfT had little idea what HS2 Ltd was doing, nor whether their actions were reasonable. Rutnam, a Cambridge University history graduate who started his career at the Treasury, wrote an article for the online employment magazine *Finito World* encouraging people to work in the civil service in 2024. He presents a long list of jobs the DfT offers: 'We had everything from policy experts through to statisticians, data scientists, social researchers, economists, lawyers, actuaries, accountants, finance experts, and specialists in estate management'.

Note the absence of any engineering roles. Network Rail, which repaired the tracks and maintained the stations, was stuffed with engineers, but they were held at arm's length by the DfT and only peripherally involved in HS2. Meanwhile TfL, which also understood engineering and had overseen Crossrail, wasn't involved much either, even in advising on the tricky London approaches.

If civil servants didn't understand how basic civil engineering works, then the MPs and the authorities tasked with scrutinising HS2 knew even less. In all the committees which looked at HS2 over the years in the House of Commons and the House of Lords, there was little discussion of the actual practicalities of building a high-speed line, nor much curiosity about the experience of other countries around the world. Thanks to Hodge and her committee, the new railway was framed early on as a project that was too expensive and too standalone to endure serious scrutiny. The transport select committee of the House of Commons, which also looked at HS2 repeatedly, was more supportive – at least in the early stages – although its more recent hearings and those in the House of Lords flushed out details of how the project went wrong, problems with contracts and the consequences of cancelling large sections of the line after 2020 and 2023.

Huge amounts of time were spent on HS2 in PAC and transport committee meetings, and in the House of Lords economics affairs committee. Scores of debates were also held in parliament itself. Government bodies, including the National Infrastructure Commission, the Infrastructure and Projects Authority, the Major Projects Authority and the National Audit Office – all of whom had overlapping responsibilities to ensure the government was spending money wisely – produced endless, sometimes conflicting reports on financing and governance.

Yet all this scrutiny often produced more heat than light. Civil servants ultimately considered that a cross-party majority wanted HS2 to go ahead and it was their task to deliver. The committee reports, right up until the present day, were obsessed with cost above all else. And their concerns weren't joined up – there's no page on the parliament.uk website that gathers together every discussion in relation to HS2, nor a gov.uk page where it is possible to access all the official reports scrutinising HS2. Each debate, report and committee occurred in a vacuum, making it impossible for members of the public or parliamentarians to connect the dots.

The fact that committee chairs were now independent and not governed by party whips – which was supposed to give them more power – meant they were sidelined when the juggernaut of government had decided to do something. They could be dismissed as troublesome mavericks.

There are real questions about whether the British political system is conducive to evaluating complex infrastructure projects like high-speed rail. Much of the discussion in parliament is adversarial, led by humanities graduates and accountants who pride themselves on winning clever arguments rather than focusing on the national interest.

Meanwhile, hundreds of millions of pounds were being spent on consultations, engineering reports and environmental studies in preparation for the actual hybrid laws which HS2 needed to proceed. Even more parliamentary time would be spent on the hybrid bill – some 1,300 hours according to the Hansard Society – making HS2 one of the most debated pieces of legislation in parliamentary history.

8

McLoughlin to the Rescue

'I was coming down to see the prime minister on the Friday when I suddenly got a phone call from Kate Fall [Cameron's deputy chief of staff] saying "could you come down and see Dave a little earlier today?"', recounts Lord Patrick McLoughlin. The parliamentary recess was just ending and the prime minister was planning his first cabinet reshuffle.

They had a drink together in Cameron's office. 'And then David sort of said, "Patrick, you know if you want to stay as chief whip you can, but I've got something else in mind for you. You can go back to the department you started your political life in" [the Department for Transport]. Because I was made a junior minister there by Mrs Thatcher in 1989.'

McLoughlin didn't say yes immediately. It was a huge change to go from a 'backroom boy' for seventeen years to a front-facing role which demanded major decisions around franchising, mending lines, a third Heathrow runway and, of course, HS2. But he thought about it overnight and then decided to make the jump. For Cameron it was a relief to have an experienced minister who would back the work of the coalition and make sure that the legislation needed for HS2 (a paving bill and then various hybrid bills) were steered through parliament.

'I didn't know much about HS2. It was a learn on the job,' he says now. But it didn't take long for him to become convinced of its merits. McLoughlin became a huge and vocal supporter in government. For civil servants working on the legislation needed for HS2, it was a relief to have what one described as 'aligning planets': the prime minister David Cameron, the chancellor George Osborne and a strong and politically experienced transport secretary. 'McLoughlin', said one civil servant, was 'up for anything'. McLoughlin, an ex-miner from Cannock in Staffordshire who represented Derbyshire Dales until 2019, believed in the necessity of investing in rail as a transport mode for the future and was happy to do whatever it took. He had a special adviser in Julian Glover, who had been seconded over from Number 10 and was a brainy transport expert on whom he could rely. Glover, also fascinated by British industrial history, subsequently wrote a book on Thomas Telford, famous for his iron bridges and road building, *Man of Iron: Thomas Telford and the Building of Britain*.

McLoughlin credits privatisation for bringing enormous private capital and (at least before the pandemic), more than doubling passenger numbers from 700 million to 1.8 billion. He loves modern stations like St Pancras which, in his words, have become 'destinations' rather than miserable places where you wouldn't want to spend more than five minutes. And he understands that much of Britain's Victorian railway structure isn't fit for purpose. He cites the Farnworth Tunnel, just a mile or so out of Bolton, as an example of shoddy Victorian building, where, when Network Rail tried to electrify the line and knock down part of the tunnels, they discovered 'all sorts of things which shouldn't have been there'. As an MP for a northern constituency, he believed in building a modern line to the North, releasing capacity on the crowded West Coast Main Line and increasing connectivity and speed.

But a battle within the Conservative Party was brewing, between the 'modernisers' (of which McLoughlin was one) and the party's predominantly Eurosceptic backwoodsmen. McLoughlin reflects on his political opponents:

> You know I love these people who tell me it doesn't matter if you get to Birmingham twenty minutes quicker, all I point out is if you get the choice between the slow train and the fast train most people will go on the fast train… I mean one of the speeches I remember making… this was my own philosophy to tell all those Eurosceptics who were so opposed to High-Speed 2. I just pointed out that I found it rather odd that I could go from London to Paris or to Brussels, or anywhere in Europe on a high-speed train but I couldn't do it to Birmingham, Manchester or Leeds.

Glover was also a strong proponent of HS2 and had discussed the idea of high-speed rail early on with Andrew Adonis. He argues that Britain can't continue to use a railway system dating back to the Victorian, or in some cases the Georgian, era – and will have to renew its railways one way or another. 'Can you imagine', he says, 'having the same lines in a hundred years' time?' He also raises a separate philosophical point, somehow never fully resolved over the last fifteen years: was HS2 to be built as a separate high-speed network to run alongside existing routes, or was it a faster part of the existing network? The TGV, he argues, was built to run on fast and slow lines, whereas the Japanese Shinkansen was a totally new and separate high-speed network.

With the minister and his adviser on board, Philip Rutnam, the permanent secretary, sprang into action. He appointed one of his most senior civil servants, David Prout, as the first director general of HS2. Prout had been director of planning at the Royal Borough of Kensington and Chelsea and had worked as John Prescott's parliamentary private secretary in the 1990s. His task was to make sure that the High Speed (Preparation) Bill, the agreement in principle for the railway to go ahead, was drafted and agreed and that the High-Speed Rail (London–West Midlands) Bill went through parliament. This second bill was the hybrid bill which would give outline planning permission for the line. Prout expanded the DfT team working on the project from fifteen to 150 to prepare the bill. At this point, civil servants in the Department for Transport had convinced themselves HS2 was not particularly high risk: Britain wasn't a 'first mover' as it might have been ten or fifteen years before. Japan, Spain, Germany and France had all proved that high-speed railways worked and China had set up a programme in 2008 which was going gangbusters. The main objective was to ensure that the legislation passed through parliament quickly and efficiently and very little consideration was given to how complex it might actually be to deliver and build. That could be easily worked out!

McLoughlin for his part sought out Sir David Higgins, an Australian businessman who had worked on the Olympics and been chair of Network Rail for two years to lead HS2, believing he was best qualified to nail down the particulars. He appointed Higgins as chair of HS2 on a £244,000 salary for a three-day week – rather more than permanent secretaries were paid for a five-day week. McLoughlin defends the decision today, arguing he wanted the best man for the job. Since then, HS2 has consistently paid its chief executive and chairs far more than other civil servants. Simon Kirby, HS2 Ltd's then CEO, was paid £750,000, the highest earning public servant in the land, a measure perhaps of how few people in the UK actually had the skills to take on the job and the high risk of failure.

At that time, cost was not uppermost in the minds of those planning HS2. Government, as one civil servant told me, always has money to spend on mega-projects and could spread the cost over decades across annual budgets. The upper echelons of the DfT considered HS2 'an act of civilisation'. McLoughlin came to believe something similar, a national project vital for UK connectivity. If you have that much faith in a project, a cost–benefit analysis becomes meaningless, just a hoop to jump through. It was in retrospect a foolish approach. Cost and the budget became the only aspect of the project the media and committees concentrated on. The other downside of this disregard for costs meant civil servants and ministers, determined to deliver the HS2 legislation come what may, were always prepared to pay off opponents and alter the engineering, often at vast expense. As in the case of Cheryl Gillan – McLoughlin remembers how ironic he thought it was that she would stand up in the House of Commons and accuse the government of overspending when it was her tunnelling demands that had driven the price up.

At this stage, many of the engineering changes were born of an understandable reluctance to disrupt other parts of the transport network. The original HS2 route out of London to West Ruislip for instance was to run alongside the Central line and, according to chief engineer Andrew McNaughton, was a perfectly viable route which should have been retained. But civil servants were concerned about disruption to the vast Hanger Lane gyratory on the North Circular in west London where bridges would have to be raised and widened. According to some it was not a direct political decision, but it was, McNaughton believes, born of a desire to retain the support of the mayor of London, Boris Johnson, who was eyeing up the Uxbridge and South Ruislip constituency, through which the line would have to run. Yet, the Northolt Tunnel which replaced this route proved expensive and tricky. Driving through the layers where the ground turned from London clay to sand, it required four tunnel boring machines, each with a price tag of between £15 and £20 million, around 160 metres long. By 2024, the completion of the tunnel was months behind schedule.

Numerous other small changes wreaked havoc with the budgets before HS2 even went to the hybrid bill committee in parliament. Spying a chance to solve long-standing problems which had been dismissed previously as too expensive, Network Rail rolled a myriad of repairs and updates into HS2, sorting out old signalling at Crewe or renewing a junction here and there.

The Treasury was not happy. Officials there were dead against HS2, despite their political master Osborne's support. They decided they were going to make life as difficult as possible, scrutinising the figures to the nth degree, as well as letting everyone who would listen know they thought HS2 was 'bollocks', as one senior Treasury official put it.

They were essentially pursing a wrecking strategy, designed to make delivering the HS2 bill as difficult as possible. Treasury officials couldn't stop HS2 – which they should have, if they were really concerned about the overruns – but made sure those in Whitehall knew their views. They too were crucial in framing the debate from the beginning about profligate spending, while distancing themselves from any accountability.

It is testimony to the power of the Treasury in the UK that they have such a hold on government. This is potentially disastrous for large infrastructure projects, which they try to control without understanding the first thing about engineering. Throughout the project, by trying to save costs, the Treasury has often added to the HS2 bill by demanding pauses and cuts which had far-reaching and expensive consequences. As someone who has knocked around Westminster said: 'The only thing they really like building is roundabouts in Guildford because they use traffic and if it moves around fast enough they'll probably pay more tax'. If the Treasury had had their way, Crossrail (the Elizabeth line) would never have been built; they suggested as much to Osborne, the incoming chancellor, in 2010. He ignored them. Since it opened in 2022, the Elizabeth line has drawn far more passengers than expected and now counts for one in every seven rail journeys in the UK.

Meanwhile, cabinet secretary Jeremy Heywood, who was panicky about the complexity, insisted the project should be managed in six-week sprints: this style of project management, popular at the time, is described as 'agile working'. Civil service guidance recommends it when 'dealing with situations with complex problems, unknown solutions and/or scope that is not clearly defined', suggesting it helps reduce costs and time. But it wasn't working with HS2 – the costs of the project rose with every sprint. Heywood told civil servants that his protégé and former colleague at Morgan Stanley, Lex Greensill, who he had controversially employed as a Cabinet Office adviser, would 'have a great way of framing it.' The suggestion went down like a bucket of cold sick. Greensill later used his influence to have several meetings with HS2 Ltd itself, presumably to 'help them' with payments to suppliers – to no avail as far as the public record shows. Greensill would also later employ David Cameron as a lobbyist for his scandal-ridden supply chain financing operation, Greensill Capital.

***

The HS2 project faced several immediate hurdles. The previous secretary of state for transport, Justine Greening, had promised to link HS2 with HS1, which terminated at St Pancras. Such a link would have made political sense, establishing HS1 and HS2 as a high-speed network, joining the North not only with London, but the European continent. It would also, at least in 2012, have made sense of the coalition government's pro-European position, cementing ties with the Continent. But retrospectively rigging up some kind of connection between St Pancras and Euston at a reasonable price wasn't easy. The more the engineers looked at the options, the more unrealistic they turned out to be. The first option was to dig two tunnels from the planned interchange in Old Oak Common. One tunnel would end up at Euston and the second would end up at St Pancras and then continue on to Paris. Unsurprisingly, that option was both expensive and, with the huge network of Tubes and tunnels under London, unworkable.

Another plan was to run a tunnel from Old Oak Common to Stratford International where it could join the line to Paris, but that would have been even more expensive, costing up to £6 billion, not to mention the fact that trains from the Continent had never stopped in Stratford, despite the 'international' designation, and new customs halls would have to be built and manned. The third option, which was included in the first iteration of the hybrid bill in 2013, was to widen the London Overground line which ran straight through Camden so that three trains an hour could run on separate tracks from there to St Pancras and then to Brussels. But the new railway tracks would cut straight through Camden Market and the bridge alterations would have entailed years of disruption to Camden and the London transport network. This solution, which was consulted on for a long time and on which transport officials were pretty keen, was finally also dismissed as not being cost-effective – although the price came in at £700 million rather than billions.

Instead, HS2 Ltd proposed what felt like the worst of all options – making HS2 passengers who wanted to board the Eurostar walk in the open air down a 750-metre 'green path' (engineer-speak for some shrubs and trees en route), lugging their suitcases through a housing estate, only to queue up at St Pancras to board a Brussels-bound train. Earlier plans for an outdoor travelator – a moving flat escalator at Euston, which HS2 Ltd liked to propose as the answer to all difficult connection problems in London, were dropped as impractical. Arup planners who wrote the final report recommended that a 'buggy service' could be provided for those who couldn't walk or needed help with baggage. The walk to St Pancras with luggage is slightly too long even in summer, but struggling that distance on foot on a cold and wet winter's day would be enough to put anyone off train travel. Arup also suggested, optimistically, that people could travel one stop on a crowded Tube, or that Crossrail 2, the now cancelled North–South Elizabeth line, which envisaged a joint St Pancras/Euston underground station, might eventually provide a covered walking route to St Pancras.

The outdoor walkway was a clunky solution, but as one senior civil servant said:

> It was a pipe dream that you could ever get on the train in Manchester and wake up a few hours later in Paris. You would have always needed to transfer onto an international line with customs and passports which would have meant segregation. That was the reason that Stratford International never became a stop for HS1 [the Channel Tunnel link]. In the end an HS1–HS2 link didn't make sense. You wouldn't have been able to run more than two trains an hour into France when the original idea was to run eighteen trains into Euston.

Other proposals which had to be dealt with were a spur to Heathrow, for which Arup also worked up plans. In the end, both Higgins and the Aviation Commission concluded an extra spur wasn't necessary. Instead, passengers from the North could change at Old Oak Common, which was to be a stop on the new Elizabeth line, or take flights from regional airports in Manchester and Birmingham which would also be joined up to HS2. As the spur would have cost £1.4 billion extra, there were sighs of relief all round. Higgins, who was behind both decisions, had proved adept at streamlining HS2 so it could go through parliament.

***

The engineering consultants around London working on HS2 in 2012 were busily changing plans, adding more tunnels and junctions and then taking them away, as news reached them from the DfT that another change was needed. Inevitably the price of the train line began to rise. Perhaps in an age of plenty no one would have cared, but this was the height of austerity. People were poorer, welfare benefits were being cut and the gap between North and South was increasing.

By 2013, before HS2 had even been voted on in principle by MPs, the government's £32.4 billion estimated HS2 budget had risen to £42.6 billion. A National Audit Office report in June of that year identified a £3.3 billion funding gap which meant the cost–benefit fell sharply from £2.60 worth of growth from every pound spent to £1.40. McLoughlin came out fighting, saying the NAO was using out-of-date figures, while the NAO retaliated, reiterating that the numbers they had used had been provided to them by the DfT.

The opponents of HS2 smelled blood. There might be a chance to stop HS2 from being passed by parliament after all.

9

Obama Tactics

Up in the Chilterns sits the picturesque village of South Heath, Buckinghamshire. A wealthy community, it is all large houses with long driveways and expansive, manicured gardens. Here, Emma Crane, her husband Tom and their friends and neighbours, Bruce Weston and Hilary Wharf, sat around a large wooden table in the Cranes' kitchen. The Cranes were both lawyers by training, while Weston and Wharf worked as transport consultants. Small talk soon gave way to the matter at hand – they were here to war-game how to stop HS2.

The train was set to pass through their village, emerging at the main crossroads near their houses. Together, they were determined to use their collective expertise to prevent HS2 from ever leaving Euston. Emma, originally from Edinburgh, was a Tory councillor. She had long been concerned with environmental issues and continues to be a green campaigner today. She was outraged – HS2 would desecrate the beautiful local countryside, upsetting valuable ecosystems and, because it would be built slap bang in the middle of her community, it would likely destroy village life, forcing small businesses to shut.

The environmental report, essential for a large-scale infrastructure project of this kind, angered them. It was 50,000 pages long and utterly inaccessible for the general reader. While the DfT saw consulting on the report a box-ticking exercise, Emma saw it as essential to hold the government to account because the line would likely displace and kill wildlife as well as destroy ancient woodland. Much of the report, she says, was 'gobbledegook' and some of it factually wrong. No one, let alone a member of the public, was going to read or understand it.

Crane was plugged into the general mood. Civilians and parliamentarians in the UK have been very worried about dramatic losses in species and habitats for years, particularly in rural areas. Intensive farming and fishing combined with climate change means the UK languishes in the bottom ten per cent of nations globally for biodiversity, with huge losses since 1970. Legislation over the last twenty years has reflected that concern. Not only have successive UK governments incorporated EU legislation into British law, but some laws have been strengthened by governments signing up to international biodiversity treaties.

Crane and her husband's ire only increased when they read the business case. 'It simply didn't make sense,' she declares. 'It was incredibly flimsy. It wouldn't have stood a chance if it had been a private sector plan.' The numbers didn't stack up and neither did the rationale.

Emma and her friends decided the only way to stop HS2 was to challenge the decision in the courts. First, she needed a team, so they turned to their village network. Many of the mums and dads at the primary school gate were eager to be part of the campaign, and many had expertise they lacked. They started to build a database, contacting every anti-HS2 group in Buckinghamshire and suggesting they unite under the HS2 Action Alliance. Later, other groups joined, some from as far afield as Cheshire, and by 2012 the Alliance comprised ninety groups from up and down the line. When I talk to her, Crane is careful to differentiate between the Alliance, which used the legal and parliamentary systems to try to sink HS2 and the direct-action techniques of the Stop HS2 campaign led by Joe Rukin, which went round the country with a blow-up nylon white elephant.

The Alliance groups were not only a source of much-needed cash; they became the core of the vocal public campaign which assisted lobbying MPs. Gillan, who had been supportive when she was a government minister, gave her full-throttle backing when she was sacked. Off the leash, Gillan employed Crane in her constituency office so she could work on the parliamentary side of the campaign. The Conservative leader of Buckinghamshire Council, Martin Tett, also sided with his local residents, opposing government plans and rallying a coalition of up to nineteen councils, the 51M group (£51 million: the cost of HS2 to each UK constituency). When the Alliance went through judicial review in 2012, they applied jointly with the 51M group and Heathrow Hub, a small grouping which was opposed to a new Heathrow spur.

This was the early 2010s, a time when lots of citizen action groups were setting up, employing the mass digital communication capabilities of the internet like 38 Degrees, an online campaigning platform named after 'the angle at which a pile of snow becomes an avalanche', as their website puts it. The government too was experimenting with e-petitions – where citizens could sign a petition for a debate in parliament. President Barack Obama had boasted about winning his first election through people power, collecting millions of small donations and encouraging local campaign groups to hold house meetings to support the Democrats' re-election. Crane said she and her husband became 'obsessed' with the success of Obama's campaign.

'We used their campaign emails and tailored them for HS2,' she said. One Obama tactic the Cranes and their campaign used was the 'money bomb' where you urge everyone to donate for your cause in the next twenty-four hours. 'We literally waited in the kitchen with some mobile phones at 8 a.m. wondering what would happen, and then they started ringing,' she said. She was amazed. People didn't pay immediately but were reminded to transfer the cash later. It was an effective way of expanding their database, raising money and asking for more regular donations.

For publicity, the Cranes' neighbours Bruce Weston and Hilary Wharf would write a report deconstructing HS2 claims and then send it to the media. Their first report concluded that sending the train to Scotland would be a waste of money because only eight per cent of Scottish journeys went south of the border and most Scots travelled within Scotland. Scottish *Newsnight* picked up the report and ran a lengthy item. Crane was exultant. The Alliance had hit upon a winning formula and began producing more reports questioning the rationale for HS2. The media, ever hungry for anti-HS2 stories, ran them prominently. The *One Show* featured an item in which local people in Buckinghamshire tried to decipher the environmental report but found it impossible even with professional speed-reading advice. And when the report went out to consultation, the government received 20,000 responses, many of them written by Crane's Alliance members.

The Alliance was increasingly confident in their assertions that HS2 was an environmentally irresponsible mega-project imposed on local people against their will and that its objectives could be fulfilled by upgrading the existing rail network. As their message was honed, the Alliance generated reams of media coverage – and neither Crane nor her team were shy about approaching the documentary makers who made *Dispatches* and *Panorama*. Other groups joined the local anti-HS2 movement, including the Chiltern Society and the TaxPayers' Alliance. Their case was compelling, because the government had failed to sell the project to people in Buckinghamshire or to consult with them in a way the population saw as legitimate. Local government leaders up and down the route from Camden to Northamptonshire felt powerless, with planning decisions completely taken out of their hands. When they raised objections, they were dismissed – perhaps not unfairly – as NIMBYs.

As HS2 was being built in a vacuum by a separate company, HS2 Ltd, civil servants didn't bother to conduct a broader discussion about the sort of transport rail links that might have benefited an area like Buckinghamshire, such as the east–west Oxford to Cambridge route via Aylesbury, which was in the planning stage and supported by the council. Instead, the department's instinct was nineteenth century: pay large interest groups off, build more tunnels to shut up others and crush the rest. They fundamentally misread the changing times – the rise of citizen power and direct action driven by the internet. But there was also disregard for the experience of other high-speed rail projects in Europe, whose architects had found that working with local government, and by extension local people and businesses, was the key to success. The 2010s was to see a fracturing of the connection between people and central government in England, between the so-called 'elites' based in London and everyone else, marked most starkly by the Brexit vote and the subsequent rise of smaller parties, including the Greens and Reform. Most of the local MPs, like David Lidington, the minister for Europe, who kept their heads down to protect their careers, found themselves swept away by Boris Johnson and the tides of history.

Emma Crane and her fellow campaigners, wanting to understand how their campaign was perceived by government, put in subject access requests. The civil servants and HS2 Ltd's discussion of Gillan was particularly scathing, mocking her clothes and opinions. MPs on her own side even made cat noises and paw swipes behind her back at Prime Minister's Questions in reference to her leopard-print outfit as she voiced concerns about HS2. Victoria Prentis, the MP for North Oxfordshire at the time, later told parliament how shocked she had been at what Gillan had found out:

> My right hon. Friend the Member for Chesham and Amersham (Dame Cheryl Gillan) had a very interesting wake-up call when she made a freedom of information request to find out what [they] felt about her personally. I have not yet grown a thick enough skin to make a freedom of information request about my name and HS2, and I know that my right hon. Friend the Member for South Northamptonshire has not, either.

The first judicial review returned a verdict in 2012. The outcome was sobering for the Alliance, with judges unwilling to interfere with a large project that was being legislated on by a democratically elected parliament. They did, however, rule that the compensation scheme was unfair and needed to be much more transparent. The Alliance appealed right up to the Supreme Court, but without success. By then the 51M group of councils had spent £1 million in the legal fight and the HS2 Action Alliance a similar sum.

Eventually, having ruled out going to the European Court, the Cranes and their lawyers went to the Aarhus Convention Compliance Committee, an international UN body of legal experts that scrutinises whether local citizens have been properly heard in environmental matters. They again received little succour. Having exhausted all methods, the campaign group decided they would instead concentrate on ensuring people received proper compensation. Crane and her family sold up in South Heath, as did their friends Hilary Wharf and Bruce Weston. The final humiliation for Wharf and Weston was when their beautiful garden was designated the ideal place for a spoil heap. Crane can't help wondering whether it might have been revenge.

***

However, the Department for Transport and HS2, as well as parts of the rail industry, were dismayed at the success of the anti-HS2 campaigns gathering speed across the country. They soon employed several communications agencies to counter the anti-HS2 arguments. They paid Stephen Frears' high-end production company Tomboy Films to produce two promotional videos worth £86,000 and hired Westbourne Communications to run their campaigns; two of their staff were seconded to HS2 Ltd to help with promotion. Westbourne also coordinated 'Yes to High-Speed Rail' and the Campaign for High-Speed Rail, described as being independent from HS2 and the DfT – though they were funded by businesses and the rail industry who had a significant financial interest in the project. They spawned lots of other 'independent' groups also run by Westbourne including Go HS2, Yorkshire Needs High-Speed Rail, North Wales Needs High-Speed Rail and HSR UK.

The most colourful lobbying campaign was run by their military-inclined chief executive, James Bethell – now restyled as the hereditary Tory peer Lord Bethell, a touchy-feely, ex-health minister focusing on wellness.

Sure, Bethell hadn't actually served in the army – his career had been split between journalism and managing the rave nightclub Ministry of Sound – but he enjoyed military metaphors: shooting down opponents, waging war on protestors, etc.

Bethell made no secret of his tactics and was happy to have a spread in *PR Week* about Westbourne's role in the Yes to High-Speed Rail campaign. A bus was bought in 2011 to travel around northern cities with 'Yes to Jobs, Yes to High-Speed Rail' emblazoned on the side. The rail minister, Simon Burns, and HS2's original promoter, Lord Adonis, came along to rally the sparse troops of Bethell's 'mini-army', consisting mainly of businesses that believed HS2 would bring more jobs to the Midlands and the North. A poster campaign – 'Their lawns or our jobs' – was also launched. There was no swell of grassroots support for HS2 in the North at that time, rather the campaign was a case of classic 'astro-turfing' – faking grassroots support – according to *Spinwatch*'s Anna Minton, an expert in lobbying who detailed Westbourne's tactics.

Bethell's approach, as laid out during a private lobbying conference (his words leaked by horrified attendees in April 2013), was not to talk about shorter journey times, but to develop a campaign which 'pitted wealthy people in the Chilterns worried about their hunting rights against working-class people in the north'. The campaign's aim was to try to tap into growing concerns of 'posh people standing in the way of working-class people getting jobs.' When an attendee asked what the point of the campaign was, he replied, 'to shit them up.' Moving on to his military metaphors, Bethell told David Grossman on the BBC Radio 4 programme *Beyond Westminster*: 'You've got to win the ground and then hold it. You can't just sit in your fortress and watch your opponents run around doing what they like. You've got to get out into the bush, using their tactics and being in their face.' A YouTube video, no longer available, shows him advising US lobbyists in 2012 that you needed to 'pick off' your critics with 'sniper scope accuracy'.

Bethell viewed people like Crane and her friends as 'insurgents' without campaign experience who needed to be quashed. The Yes to HS2 campaign and its offshoots now look remarkably like a rehearsal for the 'Yes' and 'No' EU campaigns which later divided the country. Dominic Cummings, who ran the Brexit campaign, was as keen as Bethell on military metaphors. His favourite gesture, according to a colleague, was to pull the pin from an imaginary hand grenade and then lob it over his shoulder as he was leaving the room.

Crane was horrified and felt the Alliance was being outgunned by big business, who were only interested in building HS2 to secure their slice of the billions of pounds on offer. The voices of those who wanted to protect the environment and renew the old Victorian network were drowned out. The government's case wasn't helped by the fact that their recruitment process for HS2 Ltd was noticeably opaque, a revolving door of consultants, manufacturers and railway engineers, many with seemingly vested interests. There was also an intermingling of roles and accountability between the Department for Transport, HS2 Ltd and 'independent' assessors of the programme. The so-called Challenge Panel, which was supposed to scrutinise the case for HS2 independently in 2011, involved people who had lobbied in favour of the railway, some of whom would likely benefit directly or indirectly.

Other pro-HS2 campaigners included the transport consultant Jim Steer, a charmingly self-effacing engineer, who had set up Greengauge 21 with the sole aim of promoting high-speed rail in the UK, and David Begg, chief executive of the *Transport Times*, who found themselves in the sights of the anti-HS2 campaigners. Both were influential in government circles, but there is no indication they did anything wrong except have fingers in lots of transport pies. Begg is a transport adviser and was director of the British Airports Authority and First Rail group. Steer ran his own rail and engineering consultancy group. Neither made any secret about where their sympathies lay.

Most of those in favour of HS2 would say 'so what'? These are engineers and transport experts who would help to deliver HS2. Why is it bad to lobby for a large rail project which would be good for the country, create thousands of jobs and cut carbon emissions? But many, like Crane, felt it was wrong that the government and private sector spent so much money on countering their arguments. It was as if might was winning over right. While she doesn't go as far as to suggest corruption, she certainly believed the effect of these efforts was to skew government expenditure towards a single controversial project over other transport projects.

Back in Westminster, despite McLoughlin's grip, quite a few other people were having second thoughts about HS2. By March

2013, Cheryl Gillan had forced the rail minister to reveal that £250 million had already been spent on the project since 2009. Fujitsu, of Post Office scandal fame, had spent £16.8 million designing HS2 Ltd's IT system, while the *Independent* reported that engineering firms Atkins and Arup had also already gone over budget. Atkins, appointed on a £13.3 million contract to design tunnels under the Chilterns, had burned through £14 million in just nine months. Meanwhile, the government compelled Arup to scale back the £10.2 million designs the company had submitted for Euston station because they were too expensive and grandiose. The company was already being paid double digit millions for other work on HS2.

Labour, who initially supported HS2, began to publicly question whether it was worth the candle. Peter Mandelson, a member of Gordon Brown's cabinet which had decided to go ahead with HS2 in 2010, wrote a *mea culpa* piece in the *Financial Times* in July 2013: 'I now fear HS2 could be an expensive mistake … all the parties, especially Labour, should think twice before binding themselves irrevocably to it.' He argued that all the resources for rail in the North were effectively going to commuter services in the South East and that HS2 would only benefit London. To this day Mandelson says he regrets supporting HS2, telling *The Times* podcast in November 2023 that it was the biggest policy mistake he ever supported and the country had now been left 'with a half-built white elephant'. Other Labour ministers were also beginning to wobble, most significantly the shadow chancellor Ed Balls. This sudden swing worried the government and civil servants. To pass the HS2 Preparation Bill in November, they needed a majority in the House of Commons and Lords – and that could only be achieved with Labour support.

Balls was on the horns of a dilemma. He was the MP for Morley and Outwood in South Leeds and his wife, Yvette Cooper, MP for Castleford, so he knew the difficulties of travelling east–west across the north of England and the frustration of his constituents. Balls also understood that it was vital to expand the economy outside London. To do that, he believed it was important to join up northern towns and cities to achieve what he has described as the 'agglomeration effect', which he has subsequently written about in learned Harvard papers. Basically, the agglomeration effect means that by connecting cities and towns you reap regional benefits. For instance, having a twenty-minute high-speed connection between Liverpool and Manchester, instead of a fifty-three-minute connection, would create a mini-region of much greater interest to investors. Build a high-speed line from Manchester to Hull with stops at places like Leeds and spurs to places like Newcastle and you could expand that region into what might become a northern belt (Tokyo style) where skills and knowledge could be shared. His instinct was to create a network here first, just as the Victorians had done, rather than start in London.

But this was not the proposal on the table from the coalition government and fundamentally, Balls was pro-infrastructure, alive to the risks of Britain's poor record and tendency to chop and change on big projects. He was under pressure from the Labour leader Ed Miliband, who had Lord Adonis moaning in his ear, and was overall inclined to accept HS2, despite it not being ideal. Nevertheless, he wasn't going to roll over and allow Osborne and Cameron to believe they had a free ride. According to advisers at the time, Balls was determined to do the opposite and take a more pugnacious line. In 2013, austerity was beginning to bite and there was a real possibility it might bring down the coalition.

Although many members of Labour would have wanted the party to reverse the cuts completely, Balls was not of that opinion and realised that to win the trust of the voters – who had largely accepted the austerity argument – he needed to prove that Labour wouldn't spend money like water and could take 'tough decisions'. Voting against HS2 or at least criticising it heavily at this stage could be a great way of demonstrating 'toughness', especially as the space for any more cuts had been firmly occupied by Osborne.

That's why Balls went to the Labour Party conference in October 2013 announcing that he wouldn't give 'a blank cheque' to HS2. He slammed the government for mismanaging it and losing control of costs. Using Mandelson's argument, he said 'it wasn't whether HS2 was a good or bad idea, but whether it was the best way to spend £50 billion [sic] for the future of the country'. The civil servants were in conniptions. They rallied Labour leaders in the North to support the project – including those in Birmingham, Nottingham, Manchester and Sheffield – and started working up a HS3 plan for the North. In the final debate, Louise Ellman, chair of the transport committee, and Jack Straw waded out in support of the bill. Straw declared:

> If plenty inside this House and outside think that there is an alternative to HS2, there is. But it is a worse alternative with more disruption. The second reason I support HS2 is because it will help to rebalance our economies. I have listened to some fancy arguments in this House but amongst the fanciest I have heard, listening to colleagues in the tearoom, is that if we put in this investment, somehow it is going to suck more economic activity into London. It is worth just turning that argument on its head, or as the Treasury like to say, 'Look at the counter-factual'.

If that were the case then it would be overwhelmingly an argument to reduce the capacity of the railways running north–south and for slowing up the lines. It is, frankly, simple nonsense.

In the end, Labour voted in favour of the bill, which was passed by 350 to 34. There were seventeen Tory rebels. At this stage it was an open secret that costs were spiralling out of control, despite ministers reassuring the House of Commons otherwise. That year, Balls also commissioned a policy paper from the prominent civil engineer John Armitt, who had just delivered the Olympics. The Labour policy paper proposed decoupling infrastructure projects from the short-term considerations of day-to-day politics. The paper imagined a National Infrastructure Commission which would have statutory independence from government and would be able to coordinate 'a coherent 25-to-30-year infrastructure strategy'.

While Osborne copied the commission idea after the 2015 election, and Armitt did indeed chair a new quango, the National Infrastructure Commission, its terms of reference meant it proved utterly toothless and became an agency of the Treasury. No grand infrastructure plan for the whole country has come before the House of Commons since and the commission acts mostly as an expert adviser, producing a series of online reports which gather cyber dust.

The preparation (or paving) bill had passed, agreeing in principle that parliament approved of the high-speed line, but divisive campaigning had polarised opinion. Politicians, civil servants, campaigners – everyone was taking pot shots at HS2. And it was only going to get worse.

10

The Naysayers

'HS2 is just one big punt. That would be fine if it were a project that could live up to the claims for it made by politicians of all three parties. But it can't.' This is Christian Wolmar writing in the *London Review of Books* in 2014. Wolmar is Britain's foremost railway pundit, a respected journalist and author of many books on the history of the railways; his website has 1,700 different articles on British rail. He simply couldn't see the point of the megaproject, declaring it to be a foolish endeavour, costing the country billions and bringing little benefit. His opposition to high-speed rail was, and remains, unshakeable.

From the beginning, Wolmar suspected politicians had oversold the benefits of high-speed rail in the UK. In the aforementioned article 'What's the Point of HS2?' he lays out the case in full. The country was too small for such fast-running trains – there was no need for trains with a top speed of 400 kilometres per hour (approximately 250 mph). The transport system would be better served by new, lower-speed lines joined to the existing network. He waxed lyrical about the InterCity 125 line connecting London to Edinburgh developed in the 1980s – far more sensible than grander *Continental* schemes like the TGV. Wolmar denied there was an issue with passenger capacity because there wasn't one in London. He conceded that there was a freight problem in the North – but HS2 wouldn't solve that. Nor did he agree that HS2 would be more environmentally friendly, because the government hadn't made the case for the carbon savings over a sixty-year period. HS2 wouldn't attract people out of their cars and planes on a grand scale – it wasn't designed for ordinary passengers, but the luxury business market. Curiously, he was far less interested in the idea that a high-speed network might bring prosperity to cities in the North.

Wolmar broadcast the idea that HS2 was a futile, untrustworthy and ultimately useless *grand projet* at every opportunity. It was a self-fulfilling prophecy. Naysayers like Wolmar turned the intellectual and political atmosphere against HS2, particularly on the *Guardian*-reading left. Thus, it became acceptable to end the line at Birmingham and throw a station at Euston into doubt in 2023.

Wolmar's arguments captured the public imagination because neither Gordon Brown's Labour government nor David Cameron's coalition government had made a consistent case for HS2. Andrew Adonis and Brown had originally proclaimed that HS2 was about speed and dragging the country into a European future with new railway lines rather than 'making do' with and 'mending' the old Victorian ones. Later arguments concerned freight capacity on the West Coast Main Line. Rebalancing the economy or levelling up also slipped in and out of politicians' arguments for HS2, but was never fully amplified and certainly very little of the public debate centred on the skills and jobs which HS2 would bring to the Midlands and the North. A lack of unified messaging made opposition easier.

Wolmar, focused more on transport than an industrial strategy, believed a stopping line out of London would be more sensible. His article gives considerable space to a report written by a couple of engineers, Quentin Macdonald and Colin Elliff, who argued it would be far better to run a new fast (though not high-speed) train up the M1 and spare the Chilterns. Instead of tunnelling out of London via Old Oak Common the train could go straight north to Brent Cross. An M1 route, they argued, could then stop at many more places such as Northampton and Milton Keynes. The line wouldn't need different rolling stock, wouldn't have to be straight and could be built to lower specifications because it would be slower than HS2. This suggestion for a different route was pooh-poohed by the DfT as being impractical, but it was more than that. The proposal didn't understand the point of HS2: to support the North and the Midlands. Nor did it acknowledge that the West Coast Main Line would continue to act as a stopping train – it wasn't being mothballed completely.

As I have mentioned, politicians never quite bottomed out on whether HS2 was a completely separate intercity network or a hybrid network which would join up to the existing one. There is little doubt that HS2 was conceived as a separate high-speed train network with its own signalling system, trains, track and stations, linking the great cities of the North and the Midlands. In the original plans, some trains were 'captive' and others would be able to continue running on the existing network. That was where the political mood lay: Osborne, Cameron and Andrew Adonis before them had been advocates of the economic benefits that connecting northern cities would bring to the country.

However, as the 2010s progressed, political arguments shifted away from large cities to focus more on 'left-behind' towns, particularly after the Brexit vote – and the government's megaproject was literally designed to bypass towns. The fiercest HS2 opponents were increasingly from the UK Independence Party (UKIP) and the right of the Conservative Party, who began to hold more sway in government, claiming to speak for towns. Labour also became more focused on towns than cities. The Labour MP and later secretary of state for culture, media and sport, Lisa Nandy, went so far as to set up a thinktank called the 'Centre for Towns' named in reference to a powerful existing thinktank, the 'Centre for Cities'. The fight for the red wall was a fight over towns and places where people had been 'left behind'. In response, politicians started talking about HS2 slightly differently. The transport secretary Patrick McLoughlin noted in his introduction to a report on the expansion of HS2 in 2015 that it 'will not be a separate, standalone railway. It will be a key part of our national rail network, and wider transport infrastructure'.

Wolmar was no Brexiteer, indeed he was an ardent remainer, but he inadvertently stepped into this unholy alliance. A vociferous opponent through his blog and podcast and across Britain's largest newspapers, from the *Guardian* to *The Times*, he ran article after article criticising HS2, gleefully reporting on rising costs. Again and again, Wolmar was able to say, 'I told you so', egged on by those other remainers in the London establishment – from the Treasury to bankers and journalists – who didn't really believe that northern manufacturing and productivity could ever be revived. There was little in it for them. These were men in their forties, fifties and sixties who had spent their entire careers managing deindustrialisation and betting the bank on global financial services.

There was a certain London snobbery to it. Living in a city where transport structure is so extensive and joined up that the majority of people in inner London don't own a car, they never experienced the frustrations of the slow and disjointed transport system that serves the rest of the country. While many of these men (and they were almost all men) might spend weekends admiring and even driving old steam trains, encouraging and investing in modern engineering for a northern industrial revival was, in their view, throwing good money after bad. Even when all the evidence had been stacking up from the time of the banking crisis in 2008 onwards that diversifying from London and financial services might be a good idea, they sneered at HS2. A commuter line to Birmingham? Who wants to go there? Manchester? Bollocks. Bang up the motorway in an electric car. The industrial future was AI and high tech and that could very well be managed in London, Oxford and Cambridge (ironically a nightmare to travel between on public transport). They didn't, as they perhaps should have done, think: money is cheap, but it might stop being cheap, so we should build infrastructure while interest rates and government borrowing costs are low.

Wolmar had powerful allies in the intellectual environmental world. George Monbiot, the environmental activist was, and still is, vehemently against HS2, in part because he doesn't believe the government has any business encouraging people to travel. In an early 2010 column he wrote: 'Progress is measured by the number of people in transit. Civilisation will have reached its apogee when the entire population of Manchester takes the train every day to London and the entire population of London takes the train every day to Manchester.' He would rather we had 'some peace and stillness in our lives'.

Monbiot, who has largely given up air travel, is an example of another stream of political thought espoused by some in the

Green Party and on the left (although Monbiot has adopted various political parties and causes during his life), that growth and economic activity is bad for the environment and that wealth needs to be shared around *and* reduced. Travel, even by train, he seemed to argue, was generally harmful and it would be better if everyone lived and worked in local areas, a kind of return to a pre-industrial utopia.

Monbiot's ideas form part of a long English tradition dating back to the nineteenth century, exemplified by the art historian and polymath John Ruskin. In 1887, Ruskin wrote to *The Times*, declaring railways 'the loathsomest form of devilry now extant, animated and deliberate earthquakes, destructive of all wise social habit or possible natural beauty, carriages of damned souls on the ridges of their own graves'. In the twenty-first century, some environmentalists have argued that building anything new from housing to train lines adds to carbon emissions, suggesting it would be preferable either to retrofit infrastructure or apply a 'low-build' model.

By 2023 Monbiot's argument had moved on from concern for peace and quiet and a simple life. Instead, he saw conspiracy from large corporations and the hand of neo-liberalism. He wrote in the *Guardian* that HS2 was 'a baggage train of bullshit', the product of 'clientelism' which he loosely defines as 'an exchange of favours, leading to the gross misuse of funds and the siphoning of public money into private pockets.' He suggests that HS2 was conceived in bad faith to enrich the elite, although he doesn't actually pinpoint how Adonis and other proponents could have benefitted.

While the HS2 project was clearly mismanaged by the government, it seems quite far-fetched to suggest that HS2 was conceived as a way of giving money to elite players in return for favours. Apart from a lot of hyperbole, Monbiot offers little evidence. In any case, the British construction companies building HS2 are the same ones who might have made a less controversial profit repairing schools, hospitals and housing on which he suggests public money should have been spent instead. Monbiot, who also advocates improving rail links in the North, is essentially of the 'make do and mend' lobby, so counter to the liberal Adonis's idea of building modern infrastructure. Monbiot's views provided the intellectual backbone of environmental protesters, who despised the idea of private companies violating, as they saw it, nature and the countryside.

A final influential voice in the HS2 debate was Professor Stephen Glaister, Emeritus Professor of Transport and Infrastructure at the Centre for Transport Studies, Imperial College. His 2021 report for the Institute for Government, 'HS2: Levelling up or the pursuit of an icon?' was a detailed takedown of the economic case for HS2. Like Wolmar, he easily dismisses the government's business case and contends that because HS2 is not part of a wider transport strategy it doesn't make sense. He is interested in the North, citing the National Infrastructure Commission which suggested that 'prioritising regional links, for example from Manchester to Liverpool and Leeds or Birmingham to Nottingham and Derby, has the potential to deliver the highest benefits for cities in the Midlands and the North.' But fundamentally, Glaister asks: if you had £108 billion to spend on transport, would you have spent it on HS2? His answer is a resounding no. The money could and should have been spent on regional schemes. Even today, Glaister says HS2 could be stopped and is progressing primarily because of the fallacy of sunk costs.

The views of HS2 sceptics did not fall on deaf ears. In 2012 YouGov found that 42 per cent of people were in favour of HS2, compared with 37 per cent against. However, by 2013 support had flipped, with 46 per cent of the population against and 31 per cent in favour. Currently, YouGov's HS2 support tracker shows only around 30 per cent of people are still in favour, with similar numbers against, though interestingly the largest group in favour are 18- to 24-year-olds, while the over 65s form the strongest opposition.

Few critics engaged with the economic case for northern cities, except to pit connections with London against east–west travel, as if they were competing ideas. Talk to politicians in the North now and they will say the two are fundamentally interlinked and the region needs both: a fast connection to London and fast cross-country lines which create agglomeration effects.

But powerful northern politicians weren't on the scene until 2017, the year the metro mayors of Greater Manchester and the West Midlands were first elected. They would become HS2's most credible advocates outside London, but by the time they took office, the arguments against HS2 had gathered too much steam. They never managed to create a joint body with the political or economic heft to influence national government in London. Desperate Conservative governments in London didn't care what northern Labour cities thought and, as Andy Street found out, Sunak was more than willing to throw a Conservative mayor under the train when it suited him.

None of the critics were concerned either with Scotland or Wales or any geopolitical advantage a high-speed rail service might bring to uniting the country at a time of deep division. In a fit of English exceptionalism, they didn't feel shame that what was the norm abroad couldn't be achieved on their tiny island

and weren't much interested in asking for international assistance. Indeed, as the 2010s went on, the mismanagement of HS2 and the political interference gave high-speed rail critics plenty to sink their teeth into and ample ammunition to accuse HS2 of being a government vanity project. By 2019, HS2 was entering a classic death spiral. As the price went up, so the critics grew and the government changed and paused the plan and so the price increased again.

There was little attempt to bring critics like Wolmar or Monbiot onside or persuade them that HS2 could be a national project which would, in the long-term, benefit ordinary people. HS2 became a polarising issue within the railway world. Ian Walmsley, an engineer and supporter of HS2, called Wolmar in 2017 'the Lord Haw-Haw of transport journalism' in the magazine *Modern Railways*, provoking a furious response. Could the critics have been brought onside if they had been consulted more? Probably yes, but instead, the focus had been on denigrating them, paying them off or bombarding them with 'facts' rather than engaging with arguments, because so much of HS2's route and its principles, vague as they were, had been established so early on.

Public support was muted and easily swayed away from HS2. In the Brexit decade with populism on the rise, HS2 could be painted as yet another example of the elites imposing their will on the people. It didn't help that the campaign for HS2 bore many of the hallmarks of the Remain campaign to keep the UK in the European Union, complete with the same cast of actors – Adonis, Osborne and Cameron.

The instigators of HS2 were naive. They believed if they could push the project through parliament with MPs' support everything would continue smoothly.

11

Hybrid Wars

The London to West Midlands hybrid bill, confirmed in parliament in January 2014, was the largest piece of legislation ever deposited. The papers weighed *half a tonne*. The bill itself was 440 pages long and included eleven volumes of drawings, with 'hundreds of plans and sections'. There was also a 2,888-page reference book which came in seven volumes, describing every parcel of land on the 140-mile route, with information about their owners. HS2 employed a logistics firm to pack up the papers and transport the numerous taped-up cardboard boxes to 250 places along the route for consultations in community centres and village halls. According to the Hansard Society, the House of Commons' standing orders were changed to permit the bill to be deposited electronically, which meant it could be sent out on a memory stick.

The bill, if passed, would give blanket authority to build HS2 through the English countryside and into the heavily built-up areas of London and Birmingham. Accompanying the bill was a 50,000-page environmental report detailing all the harm wildlife and the environment was likely to suffer, with suggestions for mitigation.

***

The bill was basically an immense outline planning document. Anyone who was directly affected had the right to send in objections and come to parliament to talk to the hybrid bill committee, letting them know why the bill might affect them adversely and demand changes or compensation. This was called petitioning, the most important part of the hybrid bill process, which had been used to obtain agreement for Crossrail in 2008 and HS1 in 1996. While it had worked in those cases, where there was far less organised opposition, HS2 was also far more complicated and affected a lot more people. To give an example of the scale: there were 365 petitions submitted against Crossrail to the House of Commons and 115 to the House of Lords. The first leg of HS2 to the West Midlands inspired around 3,700 petitions in total, ten times as many.

Petitions could be sent to the House of Commons for a period of just five weeks in 2014. Petitioners had to submit their objections and proposals on a printed document and deliver it to parliament by hand. Petitioners all had to include specific wording, calling themselves 'Humble Petitioners' and 'praying to the Honourable House' before signing off with: 'And your petitioner will ever pray etc...' Needless to say, it was not a process designed for the average person in the street and was later simplified for the other hybrid bills put in front of MPs for Phases 2a and 2b (the parts of the route from the West Midlands to Crewe and Crewe to Manchester).

The bill took almost two years to prepare and was only possible 'using a large part of the country's engineering and environmental resources'. The costs ran into hundreds of millions of pounds.

***

The environmental report compiled for HS2 was an important component and explained the possible damage to every tree, grassland and patch of ancient woodland which might be affected in the line's vicinity and mapped every bat, newt and owl that might be in mortal danger from a high-speed train blasting through the countryside. HS2 had employed hundreds of ecologists and environmentalists who clambered up trees searching for bats, scrambled through undergrowth to identify badgers, tested ponds and waterways for rare newts, scoured chalk meadows for flowers and butterflies and tagged barn owls to monitor how they flew.

The government and HS2 were very much alive to environmental concerns and were receiving a battering not only from residents along the route like Crane and her friends, but from organisations which would have traditionally supported the Tories and the Liberal Democrats like the Campaign to Protect Rural England and the Woodland Trust. So, the DfT had decided not only to make HS2 the most complex engineering project in the world, but also to make it the first major infrastructure project globally ever to commit to no net loss of biodiversity. HS2 not only complied strictly with EU and British law but committed to going *much* further. HS2 was to become, with all these measures, Britain's biggest environmental project.

Civil servants found their own secretary of state, McLoughlin, rather cynical about this approach. He told me:

> I got into trouble as secretary of state for transport once when I suggested to the Department for Transport that we should buy a few thousand [herons]. Because if we put herons onto a site and they eat the great crested newts, that's fine, because that's nature, but if we try and remove them [the newts] we are in serious trouble… so my answer was that we brought in a lot of herons, so they bloody eat the bloody things because they are not in short supply. I'm not against the environment, I get the environment, I love the environment, I represented one of the most beautiful constituencies in the country… but I'm also keen on connectivity between our great cities.

Other countries building high-speed rail had taken a more relaxed attitude towards the environment. The Chinese weren't too bothered about killing wildlife and displacing people – although in 2020 Chinese researchers published reports about the potentially fatal effect of high-speed rail on the crested ibis. The Spanish attitude has essentially been 'build first, ask questions later'. As a result, the government landed in hot water with the European Commission when they only agreed to give the Campiñas de Sevilla, an area full of rare birds and wildlife, protected status *after* they had started construction on the Sevilla to Almería high-speed line.

The environmental report didn't reassure the public, because as Crane had demonstrated, it was far too weighty for a normal person to read through – but the report did provide detailed ammunition for all the wildlife and woodland charities. Quite a few environmental concerns had been flagged but mitigations hadn't been fleshed out and wouldn't be until detailed designs were wrought in 2020. So, it was easy to petition parliament for maximum protections, with the hope, at least in some cases, that objections would lead to HS2 being cancelled. Conservation charities had a field day.

There were two big pinch points for HS2 at the hybrid bill stage for these charitable environmental campaigners: trees, in particular ancient woodland, and rare bats. Both were to cause HS2 time and expense and public opprobrium. The Woodland Trust and the Wildlife Trusts saw an opportunity to raise their profile and highlight the destruction HS2 was going to cause the countryside – and as the hybrid bill process moved forward identified more and more ancient woodland that might be lost. In 2013, the Woodland Trust announced that forty-three ancient woodlands were threatened with loss or damage, but as the charities carried out more 'research', numbers rose until another organisation, the Wildlife Trusts, suggested that 108 ancient woodlands would be threatened along the whole route.

To the uninformed public, it was a terrible desecration and the campaign to save the trees gained huge momentum. The Woodland Trust's chief executive Becky Speight declared that 'transport simply cannot be called "green" if it results in the destruction or damage of precious ancient woodland.' Her view was backed by Green Party MP Caroline Lucas and other environmentalists. In reality, HS2 was only going to fell thirty hectares in the first phase (out of a total of 609,990 hectares around the country), a fraction of the amount of woodland some of the charities were claiming. But trees evoke strong emotions. By the time HS2 had gone through the hybrid bill process, tree campaigners had managed to persuade government of the need for an extra few miles of multi-million-pound tunnelling to save another ten hectares of ancient woodland.

The other major concern for environmentalists during the hybrid bill process was bats, particularly rare bats. The Bechstein's bat is no ordinary bat and is protected by British law. There are only 21,000 out of a national bat population of several million. Ecologists surveying the route to prepare the environmental report unearthed a 300-strong colony in Sheephouse Wood (part of Bernwood Forest) in Buckinghamshire. They ascertained the line was to be driven straight through the trees, dividing male bats from female bats. If bats were killed by a high-speed train while trying to breed, their colony's survival would be under threat. Natural England, a quango set up by the Labour government in 2006 to enhance protections for wildlife – and to demand mitigations if wildlife was disturbed – declared they wouldn't grant a licence for HS2 unless a solution could be found to protect the bats. HS2's chief engineer Andrew McNaughton thought mitigation could be relatively simple, particularly as the area wasn't pristine countryside – there was a huge waste tip two woods along from the small bat colony at Calvert. He told me:

> We always knew they [the Bechstein bats] had to be protected... we were told by a bat specialist that bats follow hedgerows and therefore a legitimate mitigation was to make sure that there were hedgerows signalling between the two woods, and where they [the bats] crossed the railway there could be a green bridge... a bridge without a road on it.
>
> So, we build a hedgerow, and we have continuous hedgerow so the railway line is down there and the hedgerow above... that

was what the plan was and that was what was in the bill. We had about £5 million in [the budget] for bridges and nice hedges. Step forward Natural England – 'Prove that bats will always follow the hedges… otherwise we won't give you a licence.' And we ended up by the time we came out of the bill with a fence so that stupid bats didn't go across [the line] at ground level. They went up over the fence across to the other side in between the bridges. Then they [Natural England] said, 'Oh what if the bat comes down the other side?' There is no evidence that any bat has been found on the front of a train. So, we got into this 'no bat can be harmed by HS2' and the result is where we are… It's a complete waste of time, because if the bat's stupid enough… they can fly in through one of the portals [and get crushed by a train].

To be fair, MPs were as interested in bat preservation as Natural England. Various other schemes were proposed during the hybrid bill process, including a bat underpass. In 2015, Professor John Altringham, a bat specialist, gave the committee a learned presentation on bat protection schemes, but as such things had never been tried with railways, he was unable to give a definitive answer on what would work best. The Woodland Trust condemned a proposed box structure with lights to put off the bats, saying that the technique was untested and bat experts thought lights might stop bats moving around as they should. Other wildlife trusts intervened including the Bernwood Forest Bechstein's Project (part of the North Bucks Bat Group), the Bat Conservation Trust and the district authority Aylesbury Vale. Even the National Trust and the RSPB got in on the act, petitioning the hybrid bill committee about the potential damage to Bechstein's bats and the lack of sufficient mitigation.

It was eventually agreed that HS2 would work with Natural England on a structure that would allow the bats to cross safely. The meshed and concrete shed or tunnel which engineers eventually came up with was going to cost HS2 and the government an awful lot of trouble and money down the line. But in 2016 MPs were convinced that Natural England and HS2 would make sure bats were protected and proposals were nodded through the hybrid bill.

Emma Crane, Cheryl Gillan and their Buckinghamshire campaigners were determined their voices were going to be heard at the planning stage and the process was not going to be just for people who had expensive lawyers. They started up workshops to train local people and Gillan's constituency office turned into a petition depositary so she could deliver them by hand herself, sparing her constituents the journey to London. After submitting a petition, people were then invited to come to parliament to make their case before a small committee of MPs (the hybrid bill select committee), with a (then) QC from HS2 present. Only MPs whose constituencies must have no 'material interest' in the project were allowed on the committee which, like a local authority planning committee, was 'quasi-judicial'. This meant the committee had to hear and discuss on the record every 'relevant' issue. As the Hansard Society explains, they also had 'significant power over the extent and form of compensation offered, which may take a non-monetary form'. The public and local leaders petitioned on all sorts of things from bridle paths to whether their house was going to be knocked down to where bypasses and bridges would be placed. There were two sets of petitioning committees, one in the House of Commons and one in the House of Lords.

McNaughton told the transport committee in 2023 that it was 'incredibly intensive' and that parliamentarians 'were very searching.' He said: 'A lot of small, detailed changes were made in response to their consideration of something like 3,700 petitions. I do not know if it is the best route, but I do not know a better one.'

The Hansard Society reckons MPs spent at least a thousand parliamentary hours scrutinising the bill. The vast majority of that time was taken up by the House of Commons hybrid bill select committee, which sat for more than 673 hours considering individual petitions from members of the public. The House of Lords committee spent a mere 258-and-a-half hours doing the same. Every amendment big and small agreed by the hybrid bill committee required more papers and more plans by engineers. For future historians, the testimonies will undoubtedly prove a goldmine, a snapshot of Britain in the 2010s.

In one hearing, David Vick, a parish councillor from Waddesdon, appeared before the committee to argue passionately about the realignment of the A41 which was slated to go over his land and for which he wanted an underpass. He discussed the engineering, the view from his house, the worries in the village that HS2 would bring more housing to the area. A tenant farmer appeared next. He said that the most arable piece of his farm would be cut in two by HS2, but he could do nothing about it because the land was owned by the Oxford University college Corpus Christi, who hadn't corresponded with him directly and so he needed the committee's help. Reams of these witness statements are scattered across the parliamentary website. As hours of committee meetings stretched on, the MPs inevitably became bored, while petitioners grew frustrated that they weren't being listened to.

Gillan was dismissive of the long petitioning process, which went on for several years and had to be halted during the 2015 general election. She believed the committee hearings meant that HS2 was able to avoid engaging with the general public. 'Constructive engagement beforehand could have promoted a dialogue away from the Committee Room and thus speeded up the passage of the bill,' she told the House of Commons in October 2016. Although Gillan was probably wrong that dialogue would have sped up the process, she made a serious point.

One of the major subsequent criticisms of HS2 has been that the company and the government set the major route in stone too quickly with little public consultation about route options. That, one civil servant told me, was deliberate. The DfT would have blighted more homes if various routes had been considered publicly. Nor was there any wider argument tested with the public about why a high-speed network was necessary and who was likely to benefit before HS2 was announced. The government's policy was simply one of 'decide, announce, defend.'

That's not to say the very rich and powerful hadn't intervened early on to get what they wanted. The National Trust was keen that guests at its Hartwell House hotel were spared a view of the train, so lobbied to have the track shifted out of the sightline and hearing of guests strolling in the park, employing their own designers to come up with a suitable green embankment. The National Trust also intervened on behalf of Waddesdon Manor, home to the Rothschild family. The TV programme *Have I Got News For You* ran a cartoon in its opening credits in 2015 showing HS2 swerving past Waddesdon and then ploughing through an Aylesbury housing estate (which wasn't the case). But the most expensive mitigations were for Butlin's owner David Allen who lived (until his death in 2016) in the 'Georgian gem' Edgcote House – Netherfield Park in the BBC's 1995 production of *Pride and Prejudice*. The line was diverted from his home, east of his ornamental pond, but when excavations unearthed a fifteenth-century War of the Roses battleground, HS2 was routed even further away, over marshes for which a multi-million-pound 500-metre viaduct had to be built.

For most people without access to ministers, and more importantly lawyers, the hybrid bill process was the first real opportunity they had to make their case. Civil servants, lawyers and HS2 agents found themselves snarled in adversarial debate with members of the public who felt they had been left out of decision-making earlier and so brought to the hybrid bill committee all their grievances. Such a high-handed approach to industrial planning might have been fine in the more deferential 1990s or the nineteenth century when fewer people had voting rights, but it was not fit for purpose in the turbulent 2010s when the public was being offered referenda on constitutional matters like Scottish independence, parliamentary voting and EU membership. In that decade, local authorities, who were under enormous financial pressure because of Osborne's austerity policies, were trialling different methods of consultation from participatory budgeting to citizen assemblies.

An Institute for Government report (2021) suggested the government needs to take preliminary engagement much more seriously and suggests setting up a commission for public engagement, modelled on France's Commission nationale du débat public (CNDP). This had been set up in 1995 to consult the French public on the environmental impact of large infrastructure projects. Such a commission, the report argues, would give local communities more of a role in shaping infrastructure decisions from the get-go, with local people publicly airing environmental concerns earlier. The French have long experience now of what works and what doesn't. Proper engagement might also have encouraged local authorities along the route to come onside. Currently in the UK we rather complacently feel that because we have individual constituency MPs (rather than a list-based proportional representation system), quite a bit of engagement is covered. But the UK system can skew an infrastructure project massively. In retrospect it is extraordinary quite how much power Gillan wielded and quite how much tunnelling she was able to extract for the wealthy Chilterns at taxpayers' expense.

Many petitions didn't even reach the committee. Petitioners were headed off at the door in the corridors by lawyers from HS2 or referred to officials from the DfT and offered mitigation or compensation. These were known as 'corridor deals'. The transport committee was fairly sanguine about them, regarding such deals as 'inevitable' and simply encouraged HS2 to be 'more timely in its engagement with petitioners'.

Other MPs were less kind. They felt people were being pressured into agreements. They also pointed out that if such settlements could be made outside the committee room, HS2 lawyers could have offered them to their constituents before they even boarded the train for London. Gillan pointed to a case in her own constituency where Buckinghamshire County Council had been promised a haul road around Great Missenden in a corridor deal, but the pledge wasn't fulfilled. 'The nature of these corridor deals means that vital discussions are not transparent and assurances cannot be enforced. In this case, my constituents feel they are left in a very uncertain and unclear position as to HS2's intentions towards a traffic management plan that will have an enormous local impact at Great Missenden.' Later in 2018, Gillan told parliament that members of the public were still facing 'intimidation and pressure from the QCs and legal teams hustling up to people in the corridor right before their petition is heard'.

Civil servants argue that the hybrid bill process was in fact the quickest way to secure parliamentary approval without facing legal challenges and point out it only took four years for the London to West Midlands Act to be on the statute books, the same time it took the first railway in the 1830s. The West Midlands to Crewe Act, which was less complicated, was approved in 2021 and the final Crewe to Manchester bill, which was abandoned when Sunak cancelled that part of the line, would have likely received royal assent in 2024.

In terms of speed, civil servants may be right. There are a couple of other laws which can be used for seeking parliamentary approval for large projects. One is the Transport and Works Act 1992 and the other the Planning Act 2008. Although both were meant to circumvent the hybrid bill process, neither have proved particularly efficient for approving projects a lot simpler than HS2. For instance, it took ten years for the nuclear power station Sizewell C to receive permission through the Planning Act and the third runway at Heathrow ran into the sand when it went to a select committee inquiry – a right under the Planning Act – and MPs settled on a different conclusion to the government.

Meanwhile, a hybrid bill examined in such detail by parliamentarians protects infrastructure projects from legal challenges – which was why the HS2 Action Alliance's and other judicial reviews failed in most aspects. Many changes were made by the government and MPs as the bill went through parliament. A connection with HS1 was dropped by the secretary of state. More tunnelling through the Chilterns was agreed upon, the heights of viaducts were raised, roads were added and more compensation agreed. The total bill for MPs' amendments according to the National Audit Office in 2020 came to £1.2 billion.

McLoughlin believed that when the hybrid bill was enacted HS2 could go ahead without other permissions and was horrified to find out in 2024 that there were 8,276 extra planning applications, permissions and environmental licences needed. While the hybrid bill committee had decided the route in broad terms, MPs hadn't been able to agree detailed designs for the viaducts, tunnel entrances, junctions, shafts, bridges and four new stations, not least because in many cases detailed designs for those structures didn't exist. Nor could they agree every lorry route, especially in and around complicated cities like London and Birmingham.

HS2 had at least seven planning authorities to deal with, many of which were actively hostile and happy to hold the process up as much as they could. Government agencies, including ones within the department itself like Highways and Network Rail, weren't much better at granting permissions either, each protecting their own bailiwicks. And of course, there were endless environmental and wildlife obligations which needed detailed agreements too. The law was rigid enough to say that if anything decided by the act changed dramatically, planning permission for that element fell away, which was to cause problems when civil engineers were building the railway and ministers were busy changing the plan on the hoof.

There are other serious questions about whether a hybrid bill is fit for purpose. How could a small group of MPs, chosen not because of any expertise but because their constituencies weren't going to be affected by HS2, adjudicate on the details of the HS2 route? Could they really be expected to digest and understand thousands of pages of technical engineering and environmental detail? Was it a good use of their time? The hybrid bill process was interrupted by the 2015 general election when the make-up of the committee changed. Did that make for a fair process? And could MPs really mandate such huge mitigations and compensation without any reference to how much money it might add on to the already constrained budget envelope? There also didn't seem much discussion during the hybrid bill process of the trade-offs between infrastructure building and environmental protections, which so exercised McLoughlin.

There is probably no ideal way of agreeing infrastructure projects in parliament and the hybrid bill process, while far from perfect, had the merit of being rather quicker than other legal routes. The big push of ministers and civil servants up to 2017 had been passing the London to West Midlands hybrid bill. When MPs finally voted in favour, there was a sudden hiatus. The champions of HS2 had all gone: Cameron and Osborne had been swept away after losing the Brexit vote; McLoughlin had been reshuffled away by the new prime minister Theresa May in 2016; and Philip Rutnam, the permanent secretary at the

Department for Transport, moved to the Home Office where he was to have an unhappy time with the new secretary of state, Priti Patel. Even the chair of HS2, Sir David Higgins, found himself a new job as chairman of Gatwick Airport, though he stayed on an extra year at HS2 to oversee the contracting. May had little time to take an interest in high-speed rail and Labour under Jeremy Corbyn was not interested in infrastructure. The 'bandwidth of government' was taken up by trying to find a Brexit deal.

No one was left who was politically accountable and meanwhile HS2 gradually faded from the national political agenda.

12

Euston, We Have a Problem

While the hybrid bill was going through parliament, a storm was brewing over Euston station. Everyone who had assumed bringing HS2 into Euston would be as easy as bringing the Eurostar into St Pancras swiftly turned out to be wrong. It didn't help that there was a terrible station already on the site and that all the parties who needed to cooperate were at each others' throats. It was a battleground for all the anger and resentment around HS2.

The current Euston station was built in the 1960s and is effectively a squat black box attached over ramps to platforms in railway sheds. A station had first been built at Euston by the Stephensons in 1837 as a terminus for the trains from Birmingham to London. The station was very simple, with only a couple of platforms and a railway shed, but to give it a bit more glamour and grandeur, the London and Birmingham Railway Company had employed an architect, Philip Hardwick, to build a grand Doric arch at its entrance. In the 1840s, when the station was expanded as rail traffic grew, the new station was built around the arch, which formed the entrance to the great hall of the station.

The old station and the arch survived until 1962 when the government electrified the line and demolished both.

Although Network Rail described the replacement as a modern 'one-stop shop' concept where passengers could buy tickets, book sleeper and ferry services and hotel accommodation in one place, others have not been so kind. Michael Palin in the 1980 TV documentary *Great Railway Journeys – Confessions of a Trainspotter* said: 'Euston always reminds me of a giant bath. Lots of smooth slippery marble and glass surfaces so that people can be sloshed quickly and efficiently around.' While Barney Ronay described the station as 'easily, easily the worst main station in Western Europe' and claimed that using it is 'like being taken away to be machine gunned in the woods by various mobile phone and soft drinks companies'. It was built squatly, with a flat roof to allow over-station development – which never happened. For many, Euston became a symbol of post-war modernism and decline, the destruction of its original Victorian arch the final straw. For conservative-minded people, restoring a station at Euston to its former glory – complete with reconstructed arch – has been a dream for decades and HS2 was, in the government's mind, an opportunity to accomplish this.

It's true that the current station is grim. More than 4,000 trains pass through every week, heading to the North West and Wales. It's the tenth busiest station in Britain, but its design makes it feel worse than it is. When a train is delayed – a frequent occurrence – the station can be frightening and dangerous. The concourse quickly becomes overcrowded and when the train is finally announced a stampede ensues as people rush towards the narrow ramps that lead to the cramped platforms. In short, there is no mainline station in London as old-fashioned and as unpleasant to visit.

A total demolition would been an opportunity to build a safer and more pleasant terminus. But when the government and HS2 Ltd saw the costs in 2012, they baulked. The price came in at forty per cent over the budget for Euston and HS2 was already running over its estimated budget by a third. So, they told engineering consultants and architects – Arup and Grimshaw – to come up with a more modest plan which could be part of the hybrid bill due to be presented to parliament. Under this next iteration some platforms would be taken out of Euston station and some new ones built next door for HS2. This plan was so modest that when Camden's council leader Sarah Hayward saw plans, she described it as 'a shed being bolted on to an existing lean-to.'

It was a classic make do and mend solution, pleasing no one; it was one of the only times the Camden council leader and then mayor of London, Boris Johnson, agreed. Locals in Camden couldn't comprehend what was going on. Many argued for a 'double-deck' station with HS2 trains coming in a couple of levels lower than the current station. The trouble was that the ground beneath, riddled with Tube tunnels, caverns and cables, made any deep underground platform system impossible.

Then, in 2014 the chancellor, George Osborne, became fired up while visiting the emerging West Kowloon MTR station on a trip to Hong Kong. He told the *Evening Standard*: 'I'm thinking that maybe we should go for a really big redevelopment of Euston. There is a really big opportunity for jobs and for housing in the area. Let's face it, Euston is not one of the prettiest of the London stations. It was last redeveloped in the middle part of the last century.'

West Kowloon station, the southern-most terminal of the high-speed line to Guangzhou, was designed by US architect Andrew Bromberg at Aedas. When it opened in 2019, an architecture journalist described it as 'one of the most extraordinary stations in the world' and the public space surrounding it as 'an adventure in three dimensions.' The station is not in the centre of Hong Kong, but was built on reclaimed land near the harbour and is a gateway to China, a country which boasts tens of thousands of miles of high-speed lines. West Kowloon had a price tag to match – $10.95 billion – six and a half times the 'budget envelope' for Euston which was then only $1.6 billion (£1.3 billion). Osborne's desire to emulate West Kowloon was oddly competitive. The station is part of a mapped-out new neighbourhood and represents the new colonial power, China, stamping its authority over Hong Kong and effectively showing two fingers to the old colonial power, Britain.

There was certainly political desire for HS2 and, by extension, Euston station to have a defined 'look' as a prestige government project. Julian Glover, McLoughlin's adviser, cites the gloss of the Jubilee Line Extension, built in the nineties. After Osborne's intervention, Arup and Grimshaw were sent back to the drawing board. The station was still supposed to be 'world class' so they came up with a golden hooded design in 2015 which 'integrated' the old Euston station without actually knocking it down, a continuation of the shed to lean-to concept, just a little grander. The station was due to be built in two stages over ten years with the HS2 platforms all lower so that shops and restaurants could be built above. This station still didn't work for any party involved. As Mary-Ann Lewis, programme manager for Camden, said at the time: 'There's a risk that [a mainline station] could just be a tart-up of the existing station in not so many words.' Others worried east–west links across the two stations would be made even worse.

At this point, HS2 was supposed to be in charge both of the station and the redevelopment around it, but as was becoming clear, HS2 had enough problems designing a station and bringing the train into the centre of the city, with little bandwidth left for planning the area around the station. Camden with TfL and the mayor of London had come up with a Euston Area Plan which had basically been ignored. So, the DfT decided to take any development work off HS2 and tender it out to yet another company.

Meanwhile, the secretary of state for transport intervened to appoint a design board for HS2, to achieve the HS2 'look'. The chair of the board was renowned architect Sadie Morgan and the panel consisted of forty-five architects and design experts. Their remit included Euston station, but they struggled to have influence. In the 2017 design which was produced, the hood of the station is less rounded and the roof resembles the hut-like entrance to a 1980s shopping centre. To cut costs, mechanical ventilation was dropped and part of the hut was open to the elements. This was described as 'natural ventilation'. Heat would come from the glass roof. The price tag was £4 billion, the same price as the previous design. By 2018, Morgan and her team were increasingly frustrated. She told a Euston Strategic Board meeting, called by Camden Council, that there was no 'narrative or vision' for the local area to guide the architects and engineers working on Euston station. She bemoaned missed opportunities to create a new part of London, claiming that 'process was driving process' without any design champion involved in decision-making.

Euston station had become not just one, but four separate projects delivered by four separate organisations, commissioned at different times under four different briefs. One was the HS2 station project run by HS2 Ltd, the second a Euston station renewal project run by Network Rail, the third a massive over-station regeneration project, which had been let in 2018 for £40 million to Australian developer Lendlease and from which the government hoped to make money to pay for the station(!) and the fourth was Crossrail 2 (postponed a year later). All, except Crossrail 2, had been commissioned by different parts of the Department for Transport, none of whom were speaking to each other. The civil servants at the DfT, who were supposed to be in control, were wildly out of their depth.

Each project was interlinked, with no one in charge overall. Linking the two stations was especially difficult as HS2 and Network Rail were barely communicating. McNaughton remembers that his former colleagues at Network Rail ceased taking his calls. Massive over-station development on a sixty-acre site (equivalent to twenty-seven football pitches long and 250 metres wide), on which the government was relying to build thousands of homes and shops, was looking, literally, like pie in the sky. The engineering involved in building anything on top of a very busy station like Euston, let alone over railway tracks, was incredibly difficult, risky and expensive, not least because it would involve sinking massive piles fifty metres down at regular intervals to hold up the structures above. So complicated was the over-station development idea that in despair HS2 proposed to sell off some land around Euston to pay for the extra engineering. The Treasury vetoed the idea and the DfT refused to cough up any extra cash either.

There were more technical problems involved in bringing trains into the station because of major roads and bridges above ground which would have to be moved and tunnels and existing railway tracks underneath. The ambition to eventually run up to eighteen trains an hour from Manchester and Leeds (four to six is normal in Europe) required a complex system of tunnels so the trains didn't run into each other. Engineers calculated they needed three tunnels which would have to 'dive under' existing tracks to enter Euston station before threading through the narrow Euston Throat (the constricted area where lines into the station divide into platform tracks). The under-track tunnels designed to do this were so complex they had to be excavated with diggers rather than a tunnel boring machine and they had to be strong enough to prevent the tracks above collapsing, given the number of other tunnels, vaults and cables already underground. Some argued that if the number of trains could be reduced to fourteen per hour the extra tunnelling wouldn't have been necessary. However, the tunnelling had been decided in the hybrid bill, and to change the planning permissions within the bill would likely be long and costly. As of today, three excavated tunnels *are* being dug out under existing tracks, at a cost of £1.2 billion.

All these complications made it even more vital that all the parties were working together – but no one was. In fact, according to insiders, the situation was actively hostile. The various public-sector bodies were at each other's throats, each pursuing different objectives. HS2 Ltd was full of macho men who couldn't come up with a design for Euston station which pleased anyone. Network Rail had too many engineers and they had decided to be obstructive because they were angry they weren't building Euston themselves (and reckoned they could do a better job).

The female-led internal culture at Camden Council, set by new leader Georgia Gould in 2018 and a new chief executive, Jenny Rowlands, was inherently collaborative, but the vibe didn't extend outside the council, especially as both women became furious when they saw the railway men run roughshod over their residents. TfL and the mayor of London were pursuing a separate agenda around the Tube and weren't inclined to get mixed up in HS2's mess, even though they wanted their underground stations at Euston and Euston Square built out and had an interest in commercial development around the station.

The most difficult arguments according to insiders were about trees, parking and taxi spaces. HS2 and Network Rail wanted their own separate taxi ranks and parking and couldn't work together. Road diversions and lorry routes to the building sites also caused acrimony between the partners. Meanwhile, HS2 was under intense political pressure, struggling to come up with a station design even close to what ministers wanted, where the specifications were changing regularly and the Treasury was forbidding them to raise extra money.

13

Another Rethink?

Euston was in chaos and the political atmosphere in Britain was febrile by late 2018, when the engineer Allan Cook took over as chair of HS2. The government had just been found in contempt of parliament for refusing to publish the full details of the Brexit negotiations, the #MeToo movement was in full swing, the USA and China were flirting with a trade war, fifteen-year-old Greta Thunberg was 'on strike' from school and Extinction Rebellion had started a major campaign of civil disobedience attracting thousands of protestors from across the country. Theresa May was still prime minister, but by the end of the year would only just survive a vote of no confidence.

Cook was in a good mood and walking to his Westminster office on his first day in the job when his phone pinged. 'It was a friend of mine who was actually a CEO of one of the largest suppliers to HS2,' he recalled. 'And he rang me and said: "What the fuck are you doing? I've just seen the announcement. You have to be out of your mind? You do realise this is a poisoned chalice?" He said: "Why didn't you call me before you actually accepted the role?"'

ANOTHER RETHINK?

I said, 'Look as far as I'm concerned, why wouldn't I want to be part of Europe's and possibly the world's biggest, most ambitious infrastructure programme? As an engineer, I regard this as the pinnacle of my career. Why on earth would I talk to you about it? There's no chance I would do that.' And he said, 'Well, I hate to say this Allan, but you will regret this. It will be toxic and affect your reputation that you have spent the last forty years developing.'

Cook had enjoyed an illustrious career; the chair of the global engineering firm WS Atkins, coupled with decades in the aerospace industry where he had been responsible for the Eurofighter Typhoon. He was born and brought up in Sunderland and now lives in semi-retirement in Fife in Scotland in a beautiful house overlooking the Firth of Forth. He still retains a light Geordie accent and is an energetic man widely respected in the engineering and construction industry. He was driven by his northern roots and his experience of poor infrastructure and transport left him with a genuine desire to change Britain for the better. He is still admired for being one of the few HS2 leaders who has been publicly honest and upfront about the major challenges of the project.

Cook wasn't meant to land the HS2 job, but his predecessor, Terry Morgan, had been forced out just a few months into his tenure because of the vast overspends and delays at Crossrail, where he was also chair. Cook, who had been on the shortlist, was appointed hastily in his place. Unlike the previous chairs Morgan and Higgins, Cook hadn't worked at Crossrail or the Olympic Delivery Authority and had more extensive experience of the global engineering industry than of regeneration, for which

Higgins had been knighted. His independence and undoubted expertise made him both an asset and a threat to the government.

When Cook arrived, he found the HS2 board dysfunctional, the project's costs and objectives poorly understood, not to mention the chaos at Euston. HS2 Ltd was unmoored in a sea of political instability, without a clear sense of direction and a government which had lost interest because of Brexit. Four civil work contracts, worth £6.6 billion, which had been signed after the hybrid bill had passed to start enabling detailed design works for HS2. Each was made up of a joint venture between British construction firms and European rail companies and each was responsible for a different section of the line. Six months after the signing of the contracts, the building firm Carillion, which had won £1.3 billion of work, had gone bankrupt, leaving £7 billion of debt. The demise of Carillion was a national scandal, even more so when the National Audit Office admitted it had known the company was in trouble even before the government had signed with HS2! Other firms stepped in to cover, but the chaos also meant that the Notice to Proceed was delayed, so that contractors could cut costs – which were already coming in at £1 billion over budget.

Presented with such a mess, Cook's first decision was to order a 'stocktake' to understand exactly where HS2 stood. It was the action of a confident company chair, keen to have a grip on the situation, rather than a civil servant bound to the government's whims. What Cook uncovered was not pretty. The enabling and design works for HS2 were costing a lot more than had been forecast – not less. When the £55.7 million budget was set for HS2 in 2016 and the first contracts let, only 3.4 per cent of the overall assessment came from professional organisations. No one knew exactly where the track was going to be laid or indeed the exact environmental impact. Designs were incredibly immature. Once they had started working, the construction companies had also discovered pylons in the way of the route – which were extremely expensive to move – ground conditions they hadn't expected and utilities all over the place.

Cook's paper is also interesting in that it highlights for the first time the ambition of HS2, by far the most complicated high-speed rail project ever attempted in the world. International benchmarking, he wrote, showed no other high-speed network had had to deal with the problems of opening up a new city centre to city centre high-speed railway in such densely populated areas. The Delta Junction being built into Birmingham for instance is made up of thirteen viaducts taking tracks over motorways, local roads, existing railway lines, rivers and flood plains. The approaches to Euston are fiendish and require major road bridges over railway lines to be heightened. The redevelopments of Curzon Street in Birmingham, Euston station in London and Piccadilly station in Manchester were each, he emphasises, complex major projects in their own right.

Cook urged more strategic thinking across HS2 Ltd, the Department for Transport, Transport for the North, Northern Powerhouse Rail and Network Rail. He criticised the government and Treasury's benefit–cost ratio (BCR) as failing to 'reflect the full impact and benefit of a transformative programme such as HS2 that will change the way the economy works for generations to come'. The more diverse benefits during the construction period – jobs created, skills gained and a pipeline of work for decades– were missed by such a rigid definition. No account was made for international construction or rail firms expanding to the UK, nor even the wider urban regeneration around stations. Cook's argument was the same as Osborne's, that HS2 could be the foundation stone of an industrial strategy for the whole country, driving innovation and creating a new generation of twenty-first-century engineers. He also urged the government in the stocktake to look at cooperation with the private sector, local authorities and development agencies, in return for future revenues. It had taken more than ten years for anyone to articulate such a clear vision.

Cook's take on costs were far less palatable to the government. With the chief executive Mark Thurston, they estimated the price of delivering the scheme to Manchester and Leeds would be up to £78 billion in 2015 prices. When he approached transport secretary Chris Grayling to tell him what he had found out, Grayling simply said to him: 'Read my lips, it needs to be £55.7 billion.' Cook didn't know what to do. The government was, on the one hand, telling him to keep going and at the same time refusing to accept it was going to cost £23 billion more than the projected budget. Cook published his stocktake in 2019 with the headline figures in the executive summary and set up regular meetings with the DfT, Treasury and HS2 to keep them abreast of both costs and progress.

Why was the price so off? Cook is clear that the Department for Transport was the client, not HS2 Ltd, even though HS2 was wholly government-owned. But the DfT was full of civil servants and bereft of both engineers and experience, taking advice from management consultants and blissfully unaware of how complicated the plans they had agreed to were. Not only were each of the stations mega-projects, but there were huge numbers of changes that had been consented to through the hybrid bill including even more tunnelling. Underground tunnels, Cook says, cost at least seven times as much as surface railways because they need special multi-million tunnel boring machines, shafts, portals and concrete lining: sixty-five miles of twin-bore and cut and cover tunnels had been agreed across the south of England. Viaducts, too, don't come cheap at five times the cost of surface tracks. The line was also designed to run on costly concrete slabs rather than ballasts (as is common in Europe) so trains could travel at the highest speeds.

While Cook was writing his stocktake, it was still not entirely clear that HS2 would go ahead. Theresa May had resigned a month before his stocktake was published and now Boris Johnson was jostling in the wings. On the stump with Conservative Party members sceptical about the value of high-speed rail, Johnson had promised a review of the project if elected. The political uncertainty was intense and Cook is clear that for the project to succeed schedules had to be adhered to. Hold-ups and delays just added more to the price tag.

Johnson hurtled into office in July 2019 with his transport adviser Andrew Gilligan. Gilligan hated HS2 from the off, distrusted the people delivering it and was determined to sink it. A former BBC journalist, he had resigned in 2003 after the Hutton Inquiry, which had been set up to investigate the circumstances surrounding the death of government scientist David Kelly. Kelly had been named as one of Gilligan's sources for his allegation that Tony Blair and his press office had 'sexed up' a report into weapons of mass destruction, leading to the Iraq War. Since leaving the BBC, Gilligan had proved himself invaluable as a hatchet man, not afraid to weaponise investigative journalism to destroy his political enemies. Johnson used him at the

Spectator (of which he was editor) to dig up dirt to damage Ken Livingstone, which helped Johnson to win the subsequent mayoral election, after which he was given a job as cycling tsar. Civil servants and others within government understandably distrusted him.

Meanwhile, Johnson appointed Grant Shapps as the new transport secretary. Cook attended just one meeting with Shapps, despite HS2 being the biggest infrastructure project in Europe. Shapps, he tells me, showed a total lack of interest even when Cook tried to engage him about civil aviation – an industry which was allegedly one of Shapps' passions. Instead, Shapps delegated all meetings and decision-making to his rail minister Andrew Stephenson and never spoke to Cook again.

Shapps' reticence may have stemmed from not knowing whether Johnson was going to bin HS2. In fact, he was so unsure about whether HS2 would go ahead that, three months after he was appointed, he personally intervened to order HS2 to stop felling eighteen ancient woodland sites along the line. According to the then mayor of the West Midlands, Andy Street, a lot of the cabinet ministers 'sat on their hands' until they saw which way Johnson was going to jump. On the other hand, many advisers in Number 10 opposed HS2, including the Number 10 policy unit director Munira Mirza. Gilligan likely hoped the new prime minister would scrap high-speed rail completely, but he was also familiar with Johnson's affection for big infrastructure projects, even totally unrealistic ones. While mayor of London, Johnson had proposed an airport in the middle of the Thames estuary – 'Boris Island', as the press christened it. Accordingly, Johnson did not scrap HS2 but announced that he was going to review it.

ANOTHER RETHINK?

Johnson chose Doug Oakervee, a previous chair of HS2, a Crossrail veteran and engineer, to lead the review, so a positive decision was probably a foregone conclusion. Tony Travers, a professor at the London School of Economics, was also on the panel as well as Andy Street and then chair of Network Rail Peter Hendy. But there was one fly in the ointment – Lord Berkeley, a Labour peer and Rail Freight Group chair who was appointed Oakervee's deputy. Berkeley delivered for the anti-HS2 cause, issuing a separate report and becoming an expert thorn in the side of HS2 in the House of Lords.

Cook himself was worried about the review and believed it was on a knife-edge. Street thought so too and employed public relations agencies to keep the case for HS2 in the public domain, believing the battle had to be won for economic regeneration in the West Midlands and the North. The government was warned by business that if they cancelled HS2 then future infrastructure projects would cost a lot more. Ultimately, Cook credits Street and Hendy for ensuring HS2 was given the go-ahead.

Oakervee made several key recommendations. The most important one was that completing Phase 1 – the route to Birmingham – made no sense without completing the routes from the West Midlands to Crewe, Manchester and Leeds, where most of the economic benefit would be achieved. Oakervee agreed with Cook's argument that HS2 in the North was the beginning of a transport network. Other east–west networks in the North – Northern Powerhouse Rail and Midlands Rail – couldn't happen without the planned HS2 infrastructure and indeed were partly being subsidised by HS2 itself. Ploughing on with the current plans as quickly as possible was the only way forward. 'The quickest way to deliver long-distance inter-city connectivity to the Midlands and the North of England is to continue with Phase 1, and to fully commit to subsequent phases,' the report says. Perhaps the most striking part of the Oakervee report was its condemnation of the contract letting, the gold plating and over-specification of contracts.

The problem, as the report makes clear, was that the Treasury was, as Cook also found, unwilling to assume the financial risk of HS2. That meant two things. First, the contractors over-specified to meet every eventuality. They didn't want to be sued (including way into the future) if the slightest thing went wrong, especially in a project of this scale and complexity. For instance, cuttings through embankments were re-enforced and designed to a much higher specification than would normally be the case on the Continent and therefore much more expensive. Then, there were other chunks added to the price to protect themselves from a design change, for instance a cancellation, an unexpected obstacle, a rise in the price of materials, a shortage of labour, a change of specification – all of which were common with HS2, as engineers kept hitting obstacles and ministers kept changing their minds.

This way of contracting is called 'cost plus' and means the government has to cover the cost of every change, plus guaranteeing a percentage profit for the companies. It's much more expensive than giving companies a lump sum. The contracts had been divided into two stages to try to keep a handle on what was happening between design and build, with HS2 Ltd supposedly able to control and bring down pricing between the two. Oakervee recommended that before Notice to Proceed could happen, acceptable prices for Phase 2 had to be agreed. In the end, Oakervee conceded that for Phase 1 to the West Midlands, it would be difficult to renegotiate the route or the contracts, in part because the act of

ANOTHER RETHINK?

parliament had specified most major works, and that these were lessons to be learned for the routes further north. As a result of the Oakervee report, Johnson announced that HS2 would go ahead. Cabinet ministers like Sajid Javid, the chancellor of the exchequer, who hadn't been keen, miraculously changed their minds when they realised what Johnson wanted. 'After that the die was cast.'

The project's price was then generally accepted to be £100 billion in 2020 prices. Johnson blamed HS2 Ltd, complaining that the company had 'not made the task easier' and the costs 'had exploded' but added: 'The cabinet has given high-speed rail the green signal. We are going to get this done.' There was, of course, no acknowledgement of the government's role in the mess. Johnson also said that the northern parts of HS2 were to be hived off to a separate project called High-Speed North overseen by senior northern politicians which would integrate HS2 with an east–west railway and a northern network. A company was registered in that name by a civil servant and dissolved eighteen months later, as the plan fell by the wayside. Johnson also announced that Euston station was to be taken off HS2's hands.

Oakervee recommended an independent partnership board should coordinate the various Euston projects – with HS2 just one of the partners – and proposed there should be design changes to integrate both Euston stations. He also suggested costs could be cut by dropping a platform and that the HS2 station should be built all at once rather than in two stages, with Old Oak Common as a temporary station until Euston was completed.

When Johnson gave the go ahead, it was February 2020. In just over a month the country would be plunged into lockdown and all the trains in the country would grind to a halt.

Cook stayed on for another year to ensure that the building work and tunnelling started. He left a few months before his term was up. His reputation wasn't in tatters as his friend had predicted, because he had played a straight bat and avoided being snarled up in the politics – although after his departure, he found himself being blamed for not telling the government the true costs, for which he had to defend himself publicly. As one of the most senior and respected engineers in the country, Cook should have been applauded for his vision and for the transparency he sought to bring, but he was essentially cast aside. Global engineering experience, a vision and straight talking, it turned out, were not wanted. Cook still passionately believes in the full HS2 scheme and the wealth HS2 could bring to the economy, particularly in the North and the Midlands, and is disappointed his stocktake was ignored.

Cook was succeeded as chair by a civil servant, Jon Thompson, a man who had no engineering experience, and was previously the permanent secretary at the Ministry of Defence and chief executive of HMRC. Thompson's appointment demonstrated a government and civil service turned in on itself, which, when it came to HS2, simply saw balance sheets and were disinclined to listen to or negotiate with engineering experts who had a much wider understanding of what was going on in the world.

14

The Special Minister for HS2

One of the promises the prime minister Boris Johnson made when he gave HS2 the go-ahead was that he'd appoint a dedicated HS2 minister who would 'bring discipline to the company' and drive down costs. The lucky man was Andrew Stephenson, MP for Pendle, a rural slice of Lancashire. Stephenson was a Johnsonite through and through. He'd backed Johnson in the 2019 leadership campaign and was his Northwest Regional Organiser. Unlike some ministers, who sit in Whitehall quibbling over policy, Stephenson liked plunging into the thick of it. Dropping Stephenson into this role, as the chief whip later told him, was part of the new government's drive to put square pegs in square holes – a surprisingly rare strategy.

Railways ran in Stephenson's blood – both his father and grandfather had worked for British Rail. Stephenson also had the benefit, for Number 10 at least, of having little knowledge, and thereby preconceptions, of HS2. His Lancashire constituency was unaffected by the build, granting him greater objectivity.

Elected in 2010, Stephenson was part of the first generation of Conservative MPs quietly taking back swing seats in the red wall.

Having attended a comprehensive school, he was both more in touch with the working class and more local than his predecessor, the grand left-wing Labour MP Gordon Prentice, a privately educated British Canadian who had previously led a London council. Stephenson enjoys recalling that his first act of teenage rebellion was to join the Conservative Party. He told *Northern Life*: 'My mother thought I'd gone mad. She would have been happier if I'd have come home with a tattoo or piercing. For me, that was when it began and I started to think this is where I want to be.'

When the call came in February 2020, Stephenson had been enjoying the Foreign Office as Minister for Africa, a job he'd held for only a few months. He had been looking forward to touring the continent, hoping to better grasp how British diplomacy was playing out; he hadn't been able to travel the previous year with every Commons vote on a knife's edge in the hung parliament.

According to Stephenson, Johnson told him, almost offhandedly: '"You've done a brilliant job at the Foreign Office, now I want you to look after HS2." Cue, a big pause from me and then a "Really?"' Unsurprisingly, Johnson insisted with his usual bluster.

'So,' Stephenson tells me, 'I went pretty green into what was about to become the biggest infrastructure project in Europe.'

Stephenson talked to Secretary of State Grant Shapps and resisted attempts to add Crossrail to his list of jobs, insisting he was only supposed to concentrate on HS2, although he did agree to fold the Transpennine Route Upgrade and Northern Powerhouse Rail into his portfolio. After reading all the documents, including the Oakervee Review, he was convinced – the government had to proceed. 'You don't keep chopping and changing... I knew there was advocates for it in government and I knew there were detractors.' But he also knew he had the backing of the prime minister and senior cabinet members. His view at the time was: 'We need to get this under control, we need to manage costs... We need to respect communities along the line of the route.'

Stephenson oversaw several important tasks which had been on hold since the election, including telling his DfT civil servants to push ahead with the Birmingham to Crewe route. The hybrid bill had been placed on ice and some feared it was to be frozen completely. His next step was to establish a joint taskforce, to stop the blame game that had been tearing HS2 apart for years. He brought together all the key civil servants and junior ministers so no one could say they didn't know what was going on. The DfT and HS2 were in the room, a Treasury minister in the shape of Jesse Norman, Lord Agnew from the Cabinet Office, Nick Smallwood from the Infrastructure and Projects Authority, representatives from Number 10 and finally, a small army of civil servants from all the aforementioned departments. Stephenson acted as chairman and insisted on a monthly dashboard of the financial performance and a whole range of other indicators.

Every demand for contingency funding over £5 million came before the board. The bat tunnel, now a one-kilometre (0.62-mile) series of concrete arches with mesh in-between, enclosing the railway, had finally been drawn up and was one of the subjects frequently on the agenda. Even though at the time the tunnel was 'only' going to cost £25 to £30 million (it would soon rise to more than £100 million), it still seemed batshit crazy to everyone involved.

Stephenson says he tried to stop the tunnel, but was blocked: 'Unfortunately this just became a thing that went round and round, particularly with the Treasury being very upset about it, but then Defra saying the legislation wasn't going to be changed and Natural England just leaning back on the legislation… We got to a position that unless the law was going to be changed, unless Defra was going to allow us to do this, we were never going to get away from having to build this absurd structure.'

Another unwelcome guest at the meetings was Andrew Gilligan, Johnson's transport adviser, who would drop in from time to time to cause maximum mayhem. He suggested once to everyone's horror that the government should take up a suggestion from the anti-HS2 lobby group, the Taxpayers' Alliance, which had proposed an independent commissioner to scrutinise costs. Given that HS2 already had a residents' commissioner and a construction commissioner and was scrutinised regularly by the National Audit Office, the Treasury and the Public Accounts Committee, the proposal was clearly designed to destroy or at least delay progress. But because Gilligan was the PM's adviser and friend, whatever he brought to the table had to be taken seriously.

By March, despite the odd hitch, Stephenson's plans to bring discipline to HS2 were all going hunky dory. He had planned for the Notice to Proceed to be issued at the end of the month so that work could start. All he needed was prime ministerial approval. And then Covid hit like a bombshell. On 23 March 2020, Johnson announced lockdown and told everyone to stay at home. All activity in the country stopped. Almost every construction site closed immediately. Work was paused at Euston, Old Oak Common and all works south of the M25. Other sites in the Chilterns closed. There was total disarray.

The go-ahead letter Stephenson needed Johnson to sign was on his desk, working its way through the Number 10 system. With daily briefings about the Covid situation, and the attention of the prime minister and special advisers elsewhere, Stephenson and his civil servants at the DfT heard nothing. The next thing they saw was the prime minister being carted off to intensive care. The former chancellor, Sajid Javid, later told the Covid inquiry that Johnson's abrasive and powerful special adviser Dominic Cummings had, in the early days of Johnson's tenure and before his hospitalisation, 'sought to act as the prime minister in all but name,' 'tried to make all key decisions within Number 10' and, when their master was rushed to hospital and placed on a ventilator in intensive care, the special adviser had tried to mastermind a takeover.

'It was widely known that Cummings was against [HS2] and trying to knacker it,' Stephenson says. 'But I never had any direct dealings with Cummings, it was like he was there in the background, he was a malign influence, but you never saw him. He left Gilligan to do the dirty work.'

And he did: Andrew Gilligan appointed himself transport supremo, with a would-be remit of killing off HS2. He told Stephenson that Covid meant there should be another review and that it was likely the Notice to Proceed would not be signed off, or at least would be delayed until Johnson was back at his desk. Stephenson describes 'one hell of a battle' which ensued. He is still furious about Gilligan's interference. 'I think when the prime minister decided in his mind he was going ahead and when the cabinet decided they were going ahead in their minds, Gilligan should have said, "Okay, you know, I've lost this battle," and backed off.' Instead, Gilligan declared there should be a pause and that Covid changed everything, including the dynamics of travel.

Stephenson, like HS2 chairman Allan Cook, believed that every day of delay for HS2 cost the taxpayer more and with Gilligan ruling the roost he lost faith that he would ever receive a signed Notice to Proceed. But he was in for a surprise. The de facto deputy prime minister, foreign secretary Dominic Raab, who was 'designated last man standing', gripped the HS2 problem and Stephenson was delighted to receive a message towards the end of April that Raab had signed off the Notice to Proceed and agreed that HS2 contractors could start on the detailed design work.

Once the press release was issued, all hell broke loose. Stephenson says he had Cheryl Gillan on the phone 'screaming' at him. Greg Smith, the newly elected MP for Mid Buckinghamshire, who had stood on a platform of 'fighting HS2, fixing roads', rang up to yell at him soon after. Other MPs raised questions about why building should start in lockdown. And Rob Butler, the MP for Aylesbury, demanded that work be stopped after a HS2 worker coughed over an elderly man in Buckinghamshire.

Today, there is much debate over whether the Notice to Proceed should have been issued so early in the process. Few detailed designs had been worked out for the route, which meant, as we have explored in previous chapters, that costs rose and rose as more and more obstacles were discovered. If the Notice to Proceed had been issued later, those involved in driving HS2 forwards today (including the current chief executive Mark Wild) argue that the government would have had a better idea of what the build would actually involve and been able to negotiate a fixed price with the civil works contractors. The June 2025

Stewart Review into HS2 is also highly critical, making the point that planning and development is vital and that 'projects don't go wrong, they start wrong'. As it was, the government was at the mercy of the companies building HS2 who were running into frequent difficulties and kept asking for more money to overcome them. Between 2020 and 2022, the various consortia made 3,000 extra 'compensation' requests for more time and money.

The four large civil works contracts for the route from London to the West Midlands had been won by four consortia: SCS Railways (Skanska, Costain, STRABAG) given £3.3 billion to build the Euston tunnels and approaches and the Northolt tunnels; Align Joint Venture (Bouygues, Sir Robert McAlpine and VolkerFitzpatrick) given £1.6 billion to build the Chiltern tunnels and Colne Valley Viaduct; EKFB Joint Venture (Eiffage, Kier, Ferrovial Construction and BAM Nuttall) awarded £2.3 billion to build the North Portal Chiltern tunnels to Brackley and from Brackley to the south portal of Long Itchington Wood tunnel; and BBV Joint Venture (Balfour Beatty, Vinci) £4.8 billion to build the Long Itchington Wood Tunnel to Delta Junction and the Birmingham spur to West Coast Main Line tie-in at Handsacre Junction. The price of those contracts, which kept rising, was much greater than the capitalisation of any of the companies, so if those construction companies went belly up, there would be few left in the UK to build anything else and most of the thousands of small companies to which they subcontracted would be at risk too; the European rail-building companies with which the construction companies were all in consortia certainly didn't fancy any risk to their business either. The problem was not helped by the fact that the brief kept changing and that all the consortia were working on different parts of the line so that if one consortia had a hold-up, it risked messing up the timetables and budgets of the others.

Stephenson says that during Covid, the government wanted to send a positive message to the world that the UK – whose economy was about to nosedive – was open for business. It was the same philosophy that drove the 'Eat Out to Help Out' campaign and the hasty stops and starts of the later lockdowns. Once recovered from Covid, Johnson was increasingly gung-ho, announcing an 'infrastructure revolution' and a relaxation of planning laws as part of 'building back better'. There's no doubt at this stage too that the Treasury and Cabinet Office, who were in all the taskforce meetings, bore just as much responsibility as the DfT and HS2 for agreeing the Notice to Proceed, as they had all the data and made very little serious effort to stop it.

As the lockdowns and limitations on social interactions continued through 2020 and 2021, others outside government were having serious doubts about the future of trains, and particularly HS2. Thinktank researchers working remotely from home began wondering whether anyone would ever take a train again. Luke Murphy from the previously supportive Institute for Public Policy Research (IPPR) said in January 2021, 'it is right that the government should keep it [HS2] under review, assessing how long-term trends might impact on the value of the project.' Tom Burke from the environmental thinktank E3G told the BBC:

> When HS2 was first proposed it was an era when investment in public transport of all kinds was in decline, so it seemed like a good idea. The economic justification has been destroyed by

Covid. The savings in journey times which justified it won't materialise in the post-Covid remote working economy.

The Taxpayers' Alliance also chipped in, with its view that rail numbers would never recover and that the capacity argument no longer held water. Despite these doubts, in 2021, HS2 and DfT placed a £2 billion order to Hitachi Alstom for fifty-four bespoke trains, which they asked to be the fastest and most environmentally friendly in Europe. In the original plans, at least some of the trains were supposed to be 'captive' (meaning they could only run on high-speed lines), but by the time they were ordered, the DfT, in a change of strategy, decided they would all be classically compatible (i.e. could run on old conventional lines and new high-speed lines). At 400-metres long, many would have to be split in two to use normal stations because of the length of the platforms.

***

Meanwhile in Euston after a brief pause, the work started again. Living with HS2 in Camden was like living in a prison. The noise was unbearable – seeing it, smelling it, *tasting* it every day. Hundreds of diesel lorries were running through the area weekly. Nasrine Djemai by this time had grown up, gone to university, finished a master's in cultural studies at SOAS and had a job back in Camden working as a liaison officer for the council. She took all the local people's complaints to HS2 officials, who responded by saying they hadn't broken any noise thresholds, that they were working within their parameters and that they could offer 'noise insulation'. HS2's slogan was 'Being good neighbours,' Djemai notes, before wryly suggesting that HS2 would have been evicted had they been council housing tenants. The noise of workmen sinking piles into the ground for the enabling works was so loud it was hard for people to work or sleep. Later, Djemai would play a recording of the sound of a JCB digger from a local's flat to a conference of HS2 engineers. 'You could hear a pin drop,' she tells me.

The figures for train travel during lockdown were catastrophic. People didn't stop taking the train altogether but the numbers dropped by 77.7 per cent to levels last seen in 1872. Private operators were amassing huge debt, little revenue was coming in and previously profitable train companies like GWR couldn't pay the government for franchises. Franchises were replaced with Emergency Measures agreements which became (optimistically), Emergency Recovery Measures agreements and then National Rail contracts. Under the contracts, train companies were paid a management fee and the DfT called the shots, or as one rail expert said: 'told them when to shit'. As a result, the government poured an extra £10.4 billion into rail operations, bringing the total subsidy to £16.9 billion. The Treasury hoovered up the revenue from fares, while the DfT bore the extra costs; railways had been nationalised in all but name. Even freight journeys – largely dependent on moving construction materials – decreased during the pandemic. Those in the Treasury, particularly chancellor Rishi Sunak, started looking at HS2 and wondering whether the UK really needed an expensive extra train line.

The price of material and labour on HS2 was creeping up as inflation rose and building was slow. There were even problems sourcing the materials for concrete as many quarries were closed during Covid and the supply chain became unreliable. The construction companies were constantly asking for more time and money. By early 2022, Stephenson, in his six-monthly report to parliament, revealed that £1.7 billion of contingencies had been added to the price.

Although Johnson had talked about building back better, the government was pouring money into small businesses through loans and grants. Some £350 billion in loan guarantees were promised in March 2020 (more than three times the overall cost of HS2). Cafés and restaurants were propped up by the chancellor's Eat Out to Help Out initiative and billions given to individuals under the furlough scheme. As the government's balance sheets looked increasingly dire there was pressure on HS2 to offer up some major savings.

Tweaks here and there would not be enough. The most vulnerable leg of HS2 was the eastern part of the Y-shaped line to Leeds. Although houses had been compulsorily purchased along the route, most notably homes on Mexborough's Shimmer estate in Ed Miliband's constituency (to his fury), very little design had taken place. No hybrid bill had been prepared to go through parliament, nor was there a mayoral champion for the line. Civil servants had always considered Sheffield a 'nice to have' and been annoyed with the to-ing and fro-ing over the route and whether the train should stop in the city centre or at Meadowhall on the outskirts. The proposed line, which threaded through mining villages in Nottinghamshire and Yorkshire, was far from straightforward. The National Infrastructure Commission was appointed to put the nail in the coffin. In December 2020, they wrote a 'needs assessment' for the Midlands and the North, which basically said that concentrating on regional links across the North and Midlands would bring greater benefits than building out the eastern leg of HS2, which could be 'phased in'. The paper today reads like a colonial document deciding the fate of a far-off region, with little local input. But the report carried all the more weight in Westminster because the commission chair John Armitt was a known supporter of HS2. Gilligan leapt on the report as final proof that even the National Infrastructure Commission was against HS2.

In retrospect, Covid probably sealed the fate of HS2 and the move to cut the eastern leg was the first indication of a government unafraid to start trimming. Nothing happened with the commission's report for months and many pledges were made by the prime minister to continue with HS2. Then, in mid-November 2021, almost a year later, transport secretary Grant Shapps announced that the HS2 route to Leeds was cancelled. At the same time, he presented an integrated rail plan (IRP) which he said would address some of the problems with connections across the North and satisfy the need for cross-country connections. The plan, a £96 billion, thirty-year upgrade programme, went down with leaders in the North like a cup of cold sick, especially as it transpired that it would be built at the expense of Northern Powerhouse Rail (a proposed high-speed line across the northern belt). The IRP was essentially a wish-list of upgrades and electrification without a strategy. Stephenson today says it was a genuine attempt to rationalise rail schemes which had been bouncing around for years, without funding streams.

Northern leaders, who had believed Johnson's pledges that he would build HS2 and Northern Powerhouse Rail, witnessed their dreams disintegrate. In their eyes, all they were being offered, in the words of the mayor of Liverpool Steve Rotheram, was a 'cheap and nasty plan' which consisted of long overdue electrification (only thirty-seven per cent of lines in the UK are electrified); the upgrade of the Transpennine Route which had been planned (but unfunded for eight years) and a tram service in West Yorkshire.

Stephenson acknowledges today that the plan he masterminded pleased nobody. Like so many of the DfT's efforts, it was top-down and, as gathered from the negative reactions on the day ('betrayal' being the word most often used), made little effort to include the major players who had voiced their concerns in consultations. Even the Conservative chair of the transport committee, Huw Merriman, expressed his dismay. Labour, in the guise of shadow transport minister Jim McMahon, asserted that the government had sold out the North: 'We know exactly what Northern Powerhouse Rail means. They had re-promised and recommitted sixty times. That opportunity looks set to be lost.' The fiasco over the IRP – which was never implemented – was yet another example of the government's 'decide, announce, defend' approach.

Two weeks after the announcement of the cancellation of HS2 to Leeds and Sheffield in 2021, the Partygate scandal exploded with a front-page splash in the *Daily Mirror* under the headline 'Boris Party Broke Lockdown Rules'. Johnson struggled to control the waterfall of leaks and the news cycle was dominated by the narrative that there was one rule for the government and another for the people. Labour and the Liberal Democrats took hundreds of seats off the Conservatives during local elections in May the following year.

The vote of no confidence in Boris Johnson, which he narrowly won on 7 June 2022, coincided with one of the most egregious examples of political interference in HS2, demonstrating how some in government, when push came to shove, put their own political survival before country (or at least before Scotland). The thirteen-mile piece of track in Cheshire (known as the Golborne link) had been included in the Phase 2b hybrid rail bill (Crewe to Manchester) going through parliament in February 2022. The track would have ensured trains north of Crewe could go off towards Glasgow and the £3 billion needed to build the link had been priced into the HS2 project. The link also acted as a bypass for the congested Crewe to Wigan part of the West Coast Main Line. Most in the rail industry considered it vital and it was pretty simple to build through open fields.

The problem was that a stretch of the line ran straight through the MP Graham Brady's constituency of Altrincham and Sale West and he'd been campaigning for years to stop it. It cut across several farms and had annoyed a lot of his constituents. The Campaign to Protect Rural England didn't like it very much either because they claimed it tore into the green belt.

When he was elected, Brady wasn't a very significant MP or a government minister, but he became increasingly vital as chair of the 1922 Committee. This committee (set up in 1923 not 1922) is made up solely of Tory backbench MPs. It is convened as a weekly meeting and a forum for backbenchers to (confidentially) express their opinions about ministers.

Even more importantly, the 1922 Committee oversees the procedure to elect the party leader and is the one to which Tory MPs write letters if they lose confidence in the prime minister. The letters are sent to the chair, in this case Brady, and if fifteen per cent of backbench MPs send him letters saying they have no confidence in the prime minister, a vote is triggered. As Tory MPs were often trying to change prime ministers during Brady's tenure – which stretched from 2010 to 2024 – he became a pivotal player in Westminster politics. He loved the attention and relished the power. In his aptly named autobiography, *Kingmaker*, Brady describes how the job meant he was in the room with five successive Tory prime ministers. 'I was the one who watched their faces as the bad news hit them,' he wrote. 'I had seen them up close: sometimes with shoes off and their feet on the table, at others clammy, tense or even tearful.'

He enjoyed telling the media that he couldn't possibly say how many letters he had received calling for a prime minister to go, nor indeed who they were from, but still intimating that the letters were piling up on his desk. When enough letters arrived, he would then call up Number 10 to say that he needed to see the prime minister. Television cameras would be briefed for the moment he went into Downing Street.

He first oversaw a vote of no confidence in Theresa May in 2019 after Tory MPs on the right of the party became angry about her Brexit deal. Although May survived, she was mortally wounded and resigned seven months later. He went on to oversee a vote of no confidence in the next prime minister, Boris Johnson.

The HS2 hybrid bill, which would give permission for the line between Crewe and Manchester, was presented to parliament in February 2022. Brady was much exercised that the Golborne Link affecting his constituency was included. At that point he was at the height of his powers, the newspapers were full of Partygate, letters were pouring into his office and the prime minister was tottering. After much lobbying in March 2022, he personally received verbal assurances from the then secretary of state for transport, Grant Shapps, that the link was going to be cancelled and relayed this information in a letter to an angry constituent. This letter was then leaked to the *Guardian*. If it was a deliberate attempt to bounce Shapps – and it probably was – it took a couple of months to work. The scrapping of the link was finally announced by HS2 minister Stephenson only two hours before the result of the no-confidence vote on Boris Johnson (which Brady was overseeing) was known.

The timing raised many suspicions, although the government says the decision had been in the diary for weeks. Stephenson today says he was just 'reviewing' the link, but Golborne was pulled from the Crewe to Manchester hybrid bill. At the time, Stephenson pointed to the opinion of Network Rail chair Sir Peter Hendy – later to become Lord Hendy, a rail minister in Keir Starmer's government. Hendy had been tasked to write an independent review looking into connectivity in the UK. A wily operator, Hendy had suggested in his report that although the Golborne Link would provide connectivity to Scotland and relieve the West Coast Main Line there *might* be better alternatives, yet to be determined:

> The emerging evidence suggests that an alternative connection to the WCML [West Coast Main Line], for example at some point south of Preston, could offer more benefits and an opportunity to reduce journey times by two to three minutes more than the 'Golborne Link'. However, more work is required to better understand the case for and against such options.

The paragraph gave enough cover for Stephenson to put it on ice, although no alternative (read larger and more expensive) option had been worked up and publicly produced.

THE SPECIAL MINISTER FOR HS2

The rail industry was furious. The Railway Industry Association, the Rail Freight Group and the High-Speed Rail Group issued an uncharacteristically angry statement pointing out that the money had already been allocated in the plans and that this would cause a bottleneck north of Crewe on the West Coast Main Line which would 'negatively impact outcomes for passengers, decarbonisation and levelling up'. A government assessment issued at the same time was also unable to paper over the fact that losing the Golborne Link without a plan would substantially reduce the benefit–cost ratio (BCR) of that stretch of the line from 'low' to 'poor to low'. No alternative was ever put forward. As it happened, however, none of this mattered, but it was characteristic of how – even as late as 2022, after bills were published – plans *kept changing*.

In the end it wasn't Partygate that finished Johnson, but the revelation in July 2022 that he had promoted the MP Chris Pincher to deputy chief whip while knowing that complaints had been made about him sexually assaulting men. Members of Johnson's cabinet, led by chancellor Rishi Sunak and health secretary Sajid Javid, started handing in their resignations and eventually the prime minister's position became untenable, with Johnson standing down days later. Stephenson, who had stayed faithful to the prime minister, was moved from rail to being a minister without portfolio. The HS2 brief was taken on by two MPs – Trudy Harrison and then Lucy Frazer – who never stayed long enough in the job to understand what was going on. With the HS2 chair Allan Cook also on his way out, Britain's largest infrastructure project was adrift once more.

Only with the resignation of Johnson's successor Liz Truss after forty-nine days (6 September 2022 to 25 October 2022) in

October 2022 did a modicum of stability return to government when Rishi Sunak became prime minister. But by this time, interest in HS2, or even levelling up, had waned as the country reeled from Truss's disastrous budget. The war in Ukraine was taking up increasing amounts of government time and beginning to affect the cost of everything from oil to building materials. The scene was set to chop the legs off the project so that it became a bloody stump of a line from London to Birmingham. Although rail travel was reviving to almost pre-pandemic levels – leisure replacing commuting – those in favour of HS2 found themselves fighting a losing battle. Boris Johnson had been an advocate because he genuinely liked big infrastructure projects but he had been surrounded by special advisers and cabinet ministers who didn't and were sceptical about the cost. When he resigned, there was no one to take up the baton.

Stephenson, who lost the swing seat of Pendle in 2024, remains a keen advocate of HS2, believing passionately in the benefits it will bring to the country for growth. 'I don't know at what point in our history we stopped building things,' he says. 'There must have been a point when things just slowly ground to a halt and they've never really got going.'

15

A Green Revolution

During Covid, HS2 embarked on the largest single environmental programme England had ever witnessed. Not only was the Bechstein's bat tunnel approved, but digging began on 226 special new ponds for great crested newts. There was other help for wildlife. HS2's ecologists – the company having become the largest employer of ecologists in the country – fashioned extra nests for barn owls, new setts for badgers, boxes for bog-standard bats and fifteen hives for the British black bee in the Colne Valley.

Engineers also instigated a frenzy of green bridge building, designing sixteen for small mammals and insects to scamper and flutter across unharmed. Given there were only six green bridges in the UK before HS2, this was a massive shift in environmental protection. Most of the bridges were relatively modest, but at Turweston on the border of Buckinghamshire and Northamptonshire near Brackley, contractors began planning the biggest green overpass ever in the UK. The 5,940-square-metre steel and concrete viaduct makes the feted garden bridge in London look modest and is set to be ninety-nine metres wide. Grass, hedgerows and trees will all be planted on the bridge with a country lane, bridleway and footpath laid across it. When completed, HS2 claim it will be impossible to tell where the natural landscape ends and the bridge begins.

Meanwhile arborists drew up a planting schedule for seven million trees along the route from London to Birmingham, creating woodland and habitats for species the line displaced and the Forestry Commission administered a £5 million HS2 woodland fund to dole out cash to farmers within twenty-five miles of the line to create more woods and a green corridor. In 2020, some ancient woodland was 'translocated' away from the line so it wasn't lost completely. Landscape architects were employed to plan a huge chalkland meadow on the southern edge of the Chilterns and designed a Western Slopes project in the Colne Valley, creating new wetlands and wood pastures. HS2 reckoned all these mitigations would create thirty per cent more habitat than was there before.

Of course, a lot of these plans were controversial. Experts dubbed the translocation (uprooting and replanting) of ancient woodland as 'gardening' at best. David Coomes, professor of forest ecology and conservation at the University of Cambridge, told the BBC that translocation 'is like tearing up a Turner masterpiece and tossing little bits of it into a new art installation and hoping people don't notice the difference', with real scepticism about whether the insects, fungi and bats will follow the replanted woodland. It's a controversial practice, but HS2 still issues an annual report on how translocation is progressing in the hope the expertise the company gathers will help future infrastructure projects.

You might have expected, post-Brexit, for biodiversity protections to have been relaxed, but Boris Johnson promised that when the UK left the EU, the country would 'build back beaver' and increase protections. Just at the moment HS2 was putting spades in the ground, the Environment Act of 2021 of which he was the architect became law, introducing stringent environmental laws including 'biodiversity net gain' requirements on construction projects. Not only did developers have to ensure biodiversity remained the same, but they actually had to increase it, which was why Defra officials were so emboldened in their support of the bat tunnel. As Andrew Stephenson explained:

> The overriding commitment the prime minister had... was that we weren't going to water down our environmental standards in any respect, because many people on the Brexit side of the argument argue that actually being able to end the export of live animals and various other things you weren't allowed to do under EU law, we were actually going to enhance our environmental standards [by leaving] so they kind of wanted that narrative or look, and say not only have we not watered standards down we've gone further than the EU on a range of things.

Other government bodies like the Environment Agency (EA), which lies within Defra, had their sights trained on HS2, ensuring that works along the line protected the water table and didn't increase flooding risk. Later, officials there spent more than £100,000 taking HS2 contractors to court over whether they were digging out earthworks properly at a sensitive area in Warwickshire. The clearly shocked judge, who eventually dismissed the EA's injunction in 2024, told both parties 'ultimately funded by the taxpayers' to 'cooperate in seeking an expedition resolution through the arbitral process.'

The measures HS2 was taking calmed most of the big wildlife and environmental charities. By 2021 many were actively giving advice to the government and HS2. As the more conservative wildlife and conservation charities became co-opted, other direct action environmental groups sprang up to challenge HS2 on the much wider grounds of increasing carbon emissions. A report by HS2 itself admitted that over its estimated 120-year lifetime the project would increase emissions by around one per cent. The HS2 trains themselves are electric (unlike sixty-two per cent of the current railway network) and so potentially clean, but emissions come mainly from carbon emitted by construction, including manufacturing large amounts of concrete and steel as well as the heavy diesel-run digging machines used to move tonnes of earth.

HS2 wasn't doing anything like the environmental damage of fracking or drilling oil in the North Sea or even nuclear power, but as politicians hadn't sold high-speed rail as a way to discourage car and plane use, it could be characterised as a project which damaged the environment. Indeed, Friends of the Earth are most critical of the 'opportunity cost' of investing in HS2 and not in other low carbon transport infrastructure like electric buses, cycle lanes and the old train network. In 2020, the BBC presenter and conservationist Chris Packham raised £100,000 to judicially review the decision to build HS2 on the basis that there had not been enough environmental assessments of the carbon impact. He was told in no uncertain terms by the high court that his review would fail and that all the environmental issues had been aired during the parliamentary process.

The Covid lockdown provided the perfect conditions for direct action against the increasingly visible works of the railway line. The HS2 branch of Extinction Rebellion called for the high-speed line to be cancelled and the money to go to personal protective equipment and the NHS. Protest camps started to spring up along the line, inspired by both the Extinction Rebellion movement and the Occupy protests earlier in the 2010s. The first camp was at Jones' Hill Wood in Buckinghamshire in 2020. When that was cleared by bailiffs, protestors moved on to Wendover where they started to build tunnels – and then to the square outside Euston station to protect trees due to be felled for a temporary car park.

In February 2021, while most of the population was under lockdown watching Netflix box sets, anti-HS2 activists secretly dug thirty metres of tunnels under Euston. They were joined by 'Swampy' – the now fifty-one-year-old environmental activist and forester Dan Hooper – who first pioneered tunnelling in the 1990s to hold up road-building works. Although a guidebook to digging tunnels is available on the internet (*Disco Dave's Tunnel Guide*, the definitive guide to tunnelling – how to do it and why), Swampy is known as the best DIY digger in the country. He was invited to Euston by one of the young protestors' parents, worried about their safety, and was also actively involved in Wendover.

Allan Cook looks back at that part of his chairman's job as being the most terrifying, not least because of the risk of protestors, or the people evicting them, dying on his watch in a collapsing tunnel. In Euston the shafts had been dug in secret and no one outside the group of protestors knew where they ran. HS2 employed specialist teams to empty the tunnels, but it took a month to clear them.

Swampy came to Euston with his sixteen-year-old son Rory. He told the press:

> I'm prepared to put my life on the line if need be. As far as I'm concerned, these tunnels are safe but we're prepared to last this

out for as long as it takes. We're in a climate emergency and the government is pressing ahead with the HS2 project that will deforest large areas, it's madness.

Meanwhile, the tunnellers were trying to apply to the courts to stop HS2 evicting them and HS2 was hiring the most senior barristers in a cross application to ensure they were removed – as well as trying to legally force the protestors to provide information about the extent of the tunnels. The cost to HS2 of the Euston protest was £3.5 million.

At the end of March 2021, activists arrived in Staffordshire and set up in the woods at Cash's Pit. They called themselves the Bluebell Wood protestors. Most of them were not local, but rather professional campaigners who saw themselves as environmentalists protecting woodland and forests. A group, many of them young men, started to build huts and treehouses in the woods. The protest, which took place near the village of Swynnerton, followed the playbook of the Euston and Wendover protests. For many in these rather staid local villages where very little happens, they were a novelty. Locals brought the protestors food. In December 2021, the Bluebell Wood encampment held a 'family day' for residents. Visitors, according to the breathless *Stoke Sentinel* report, were able to gather around a campfire, make natural wreaths and hear from speakers about their campaigning experiences.

For HS2, however, they were threatening to disrupt operations completely. Work had not started on this part of the route, but preliminary investigations were taking place. Boreholes had been dug; surveys carried out. Protestors were not only occupying land that HS2 needed for the railway line but blocking the entrance to sites where HS2 works were taking place and trying to take equipment. There's a video of protestors posted to their Facebook page showing them in a field next to a digger which they 'intend to board'. They are kept off by a group of around twenty contractors and security personnel. The protestor comments on his livestream without irony: 'They've had to pay for a very expensive human fence to be put here today. You know what your taxpayers' money is being spent on.'

In May 2022, HS2 issued the camp with an eviction notice. It was then the protestors started to occupy the tunnels they had been building under the encampment and refused to leave. Some stayed underground for forty-six days. A young woman on Facebook, equipped with a head torch and mobile phone, describes how she brought food and other supplies down so that she could survive as long as possible.

It took six weeks to clear the Bluebell Wood tunnels. As they sealed them off, HS2 had to bring in specialist air monitoring machines and pumps to provide air, because of their fears of poisoning protestors with carbon monoxide when blocking off sections of the tunnel to extract them. Even after everyone seemed to have gone, the company felt obliged to launch a search and rescue operation for three weeks for fear that someone had been left behind. The estimated cost to HS2 for recovering the land came to a total of £8.5 million.

The tunnelling protestors had a profound effect on the way HS2 and the government operated and ultimately may have shaped HS2's secretive and defensive culture, which issues former staff with non-disclosure agreements and even imposed them on local councils. It's impossible to find out, for instance, all the companies subcontracted by HS2, or when some works are starting, as freedom of information requests to the company are often refused on the grounds that releasing it could lead to criminal activity.

The reason protestors build tunnels is that to clear them is both expensive and dangerous. It holds up projects and protestors say it can end up with contractors considering it too much trouble to continue work on a project. Treehouses and woodland encampments are easy for professional bailiffs to destroy. 'Lock-ons', common forms of English protests since the suffragettes, are mainly good for garnering publicity. In Euston a local vicar and her artist parishioner gained national publicity by chaining themselves to a tree which was about to be felled outside the station, though she rather spoilt the effect by saying it was a 'symbolic act' and she didn't intend to hold up construction work.

The Department for Transport and the Conservative government became very pre-occupied with who the protestors were and how to stop them. In 2022, HS2 secured a high court injunction for the whole route from London to the West Midlands. As part of further injunction proceedings for the second part of the route from the West Midlands to Crewe, James Dobson, a specialist security consultant and adviser for HS2, provided details of some of the protestors in Staffordshire. Many named people who had been part of the encampment had their injunctions dropped because they had pulled out of these activities and gone back to their lives. But Dobson is keen to demonstrate that the hardcore activists operate all over the country on different campaigns.

The Bluebell Woods protest leader was Ross Monaghan. He's described as the person who established the camp in Swynnerton and 'participated in its fortification to hold out against eviction', although it's also noted that he 'left before the enforcement operation began'. He has the pseudonyms 'Squirrel', 'Ash Tree' and 'Lock Pick'. Dobson writes that before Monaghan started campaigning against HS2 he actively campaigned against fracking and suppliers of the onshore gas and oil industry. He had a criminal record from other HS2 protests – in 2021 for assaulting two security guards and for criminal damage at the Jones' Hill Wood protest site near Wendover in Buckinghamshire on the Chilterns stretch of the line. Monaghan had also been part of the Stonehenge Heritage Action Group camp – preventing a tunnel being built under Stonehenge – and had 'actively scouted land yet to be possessed on Phase 2a and Phase 2b of HS2'. He is described as someone who, since 2017, had 'returned to activism on multiple occasions.'

Others like James Taylor (aka 'Jim Knaggs', 'Run Away Jim' and 'Tim Blagg') and Leah Oldfield (aka 'Lou Pole') are cited as being activists at multiple locations. Dobson writes that 'serial campaigners' take direct action for Palestine Action, Insulate Britain and Just Stop Oil. Several targeted Kier, the construction company building a mega-prison contract in Sutton, and the law firm Eversheds Sutherland who advised on HS2's northern leg, specifically around injunctions.

Many campaigners went by nicknames, but were public about their activities, chronicling them on Facebook and HS2 Extinction websites – including comments, videos and calls to action for other campaigners, effectively making it easy for HS2 and their security consultants to track their activities. The Bluebell Wood activists didn't achieve as much publicity as those in London. Like a wave of recent protestors around the world, they were not ready to negotiate or be paid off. They were simply convinced HS2 was as environmentally damaging as roads and oil production and wanted the railway stopped. On Facebook, campaigner Connor Nichols (aka

'GoldiLocks'), who lived in encampments up and down the line, explained his motive for taking part in both the HS2 campaign and the Stonehenge protest: 'We all want an end to violence against mother nature from the state and from the patriarchy'.

One of the effects of these high-profile direct-action campaigns was an increasing determination from Boris Johnson and Rishi Sunak's governments to clamp down on protest, particularly direct action. They wanted more legal deterrents for the likes of Swampy and his son. The main crime that protestors at Euston and in Staffordshire could be prosecuted for was aggravated trespass. Some were taken to court for criminal damage and assault, but there were few other laws the state could use against them. And they were frequently let off by judges, who concluded they were exercising their right to free speech. The Euston protestors were acquitted at the first hearing in 2021 because no construction work had started on the site so it was argued no crime had been committed. They were prosecuted a second time after the Crown Prosecution Service applied to the High Court to re-try them. The six protestors were found guilty in 2022 of 'aggravated trespass' and sentenced to between one and three months in prison, suspended for twelve months.

Four Bluebell Wood tunnellers were prosecuted at around the same time at Birmingham Crown Court for breaching injunctions issued while they were in the tunnel forbidding them to be on HS2 land. They all received prison sentences of less than a year and two had their sentences suspended. Among those jailed was Swampy's son Rory, who had escaped prosecution for his activities at Euston.

It wasn't just HS2 of course. Just Stop Oil activists were holding up motorways and Extinction Rebellion was upping its activities with some protestors gluing themselves to Tube trains during rush hour. But some of the provisions of the Public Order Act, given royal assent in 2023 and considered by Amnesty International and others to be the most draconian anti-protest laws ever in the UK, seem clearly directed at HS2 protestors. It is now a criminal offence to cause 'serious disruption' by tunnelling or being present in a tunnel. Locking on – tying or gluing yourself to a person or an object – has also been criminalised, as has obstructing major transport works or interfering with key national infrastructure.

In its written submission to the House of Commons bill committee, which scrutinised the Public Order Act legislation, HS2 alleged that this law was needed because protestors had caused £126 million worth of damage. The final figure included not only the cost of removing them, but also security fencing, eviction operations and the knock-on cost to the programme from delays.

Ironically, the act has so far not been used against HS2 protestors but against Just Stop Oil demonstrators, including one who was jailed for walking slowly down a road in Holloway in north London, breaching Section 7 of the act because their actions were considered 'interference with the use or operation of key national infrastructure'.

No one is a winner here. The Stop HS2 protestors with their encampments can celebrate that HS2 will no longer run north of Birmingham, although it wasn't their action which stopped it. The government has managed to criminalise and frighten off campaigners who want to cause disruption to railway building, though at a huge cost to freedom of expression. Meanwhile, HS2 has an excuse to carry on letting out contracts and renting out land without being held publicly accountable, on the grounds that protestors might otherwise disrupt their commercial activities.

16

'Robbing the White Elephant to Pay the Red Wall'

With Johnson gone and Liz Truss consigned to the dustbin of history after forty-nine days as prime minister, Andrew Gilligan ratcheted up his campaign against HS2. No longer at Downing Street, he stepped out of the shadows and went public. Approaching his friends at Policy Exchange, a right-wing think-tank, he masterminded a paper suggesting that scaling back HS2 would be a simple, trouble-free way of saving billions over the next decade or two and help rescue the economy.

By late 2022, the Ukraine war was in full swing and the country was in the grip of an economic crisis as food and energy prices skyrocketed – people were scared. By cutting taxes and raising borrowing, Truss had caused panic in the bond market, revealing fundamental weaknesses in the British economy. Conservative MPs had thrown her out and replaced her with ex-chancellor Rishi Sunak, hoping that as a money man, he would know how to deal with the financial crisis.

In his government's first budget, Sunak indicated that he would achieve £54 billion of fiscal tightening, including

£33 billion in public spending cuts. Released in budget week, Gilligan's paper 'The Kindest Cut of All' suggested that 'Scaling back HS2 could alone deliver almost a tenth of the spending cuts required, £3 billion per year by 2027/28, significantly more (up to £7 billion per year) in later years and perhaps £44 billion or more in total.'

Usually, thinktank policy papers like this gather dust on the shelves, but Policy Exchange was a different beast. The thinktank was run by the already ennobled Dean Godson, a former *Daily Telegraph* journalist and intellectual manqué, and chaired by Michael Gove, then levelling up secretary. It had established itself as an important player in Westminster, informing Conservative Party policy and replacing the moribund Conservative Research Department (CRD). Where once it had been important to have passed through the hands of the CRD to become a special adviser, MP or even minister, by the mid 2010s, it was vital to have either worked for, or at the very least written a paper for, Policy Exchange. In 2014, Sunak, himself a protégé of Godson's, was briefly head of Policy Exchange's Black and Minority Ethnic (BME) Research Unit.

Gilligan's main objection to HS2 was the cost. 'HS2 is Britain's greatest infrastructure mistake in half a century,' he proclaimed with typical hyperbole. 'Even at the official price, even before the spending crisis, and even before Covid, it was and is a misdirection of resources of unprecedented size.' In contrast to other reports, including Oakervee's three years before, Gilligan argued that most of the benefit would accrue to London and the South East. He quoted the National Infrastructure Commission's report about how improving east–west connections would bring the most benefit to the North. Cancelling the leg to

Sheffield and Leeds had left the general public unmoved, he argued, opposed only by 'the elites' – by which he presumably meant Lord Adonis, although the newly elected mayor of West Yorkshire, Tracy Brabin, and the leader of Leeds Council had also objected.

Meanwhile, he wrote that Manchester wasn't keen on HS2 either. To be fair, the city council petitioned against the final Manchester leg of HS2 because the city wanted a through station below ground at Manchester Piccadilly, but the leader of the council and mayor of Greater Manchester were fundamentally HS2 supporters. In retrospect, perhaps, threatening opposition to HS2 was a reckless approach and demonstrated how out of touch northern politicians were with the political mood in London and the adversarial culture around HS2.

Gilligan argued there were few risks to shortening HS2 and the money saved could go to northern seats which the Conservatives wanted to retain. Ben Houchen, the Tees Valley mayor, then darling of the Tory party, had said that if he were given even one per cent of the HS2 budget he would be able to revolutionise his area's public transport on a scale we couldn't possibly imagine today. Gilligan quoted ConservativeHome's idea that it would simply be 'robbing the white elephant to pay the Red Wall.'

This idea was gaining ground within the Conservative Party. As we have seen, Graham Brady, chair of the 1922 Committee and MP for Altrincham and Sale West, had already persuaded Johnson and Shapps to cancel the Golborne Link. His faction – an anti-George Osborne, pro-town lobby – was now firmly in the ascendant and hungry for more change. Their big idea for helping the Conservative-inclined North – Brexit – was unravelling and now that Brussels was no longer on the political chess board, they needed another 'elite' enemy at which to tilt. HS2 was the perfect dragon to slay.

For Sunak, who would face a general election in under a year's time, cancelling HS2 was a seductive idea. He wanted red meat to throw at MPs who were being challenged by an ever-popular Reform Party, successor to the Brexit Party. Even without Nigel Farage, who had resigned in 2021, Reform was polling at ten per cent in early 2023. Sunak had to prove that, despite a career in investment banking and his marriage to the Indian heiress Akshata Murty, *he* wasn't part of the hated elite who needed to be culled. More practically, he could scale back a mega-project which the government had lost control of and redistribute the money to more manageable local transport schemes – and mending potholes in seats the Conservative Party needed to retain.

There were other practical reasons for cutting the country's capital spending. The economy, which had never fully recovered from the 2008 banking crisis, hadn't improved when Sunak took over. Inflation was still rising around the world after Covid, and this was particularly marked in the UK, still suffering shockwaves from Brexit. In May 2023, the inflation rate peaked at 6.5 per cent, the highest it had been since 1991, and the price of borrowing soared. Construction costs had risen by a quarter and for HS2 that meant paying up to twenty-five per cent more for concrete, timber, steel aggregates, labour and energy. HS2 was vulnerable, not only because it was the biggest capital project in the country, but because it had become a byword for financial mismanagement. If it had been toxic in 2018, it was positively radioactive by 2023.

In March, just as Sunak took on Andrew Gilligan as transport adviser, the secretary of state for transport, Mark Harper, had announced that he was 'pausing Euston' until an affordable solution could be found for the station. The Treasury had long been suspicious that costs at Euston were out of control. The year before, 2022, the National Audit Office had swooped in again and were scathing. They found HS2 had already spent £548 million preparing for the station without having agreed a final plan. This included £105.6 million on four designs by Arup, Grimshaw and WSP, all of which were subsequently scrapped, a colossal waste of government money. HS2 had also spent £1.5 billion on the approach to Euston station, which still hadn't been fully resolved.

The NAO didn't believe the Euston Partnership, set up after Oakervee to oversee everything, and chaired by Peter Hendy (Lord Hendy, the rail minister appointed by the Labour government in May 2024), could be effective because the three Euston projects (Network Rail's, Lendlease's and HS2's) were at different stages and had different business plans. The auditors also determined the £2.7 billion budget for the HS2 station as totally inadequate with designs now coming in at £4.8 billion and above (at 2019 prices).

For the very first time, the report also highlighted the cost of over-station development. All the hold-ups (and presumably fights) meant that 'the overall scope of the project remained uncertain until late 2021'. Yet, the NAO report warned that a pause such as the one Harper had ordered added further risk, and might lead to extra costs, due to stopping and starting, contractual changes (which are always expensive because construction companies demand a lot of compensation) and the business of managing the site for longer with the schedule totally out of kilter. Even before Harper ordered the pause, Euston wasn't expected to open until 2040 – and HS2 had already downgraded its station objectives from 'world class' to 'highly functional and affordable.'

Harper also decided to rephase the second chunk of HS2 between Birmingham, Manchester and Crewe by two years, meaning the track to Crewe was unlikely to be built until 2036 and the line to Manchester by 2043. Harper, who was trying to stave off the prime minister, also announced that the signalling at Crewe was to be a joint venture between HS2 and Network Rail and was to be awarded an extra £1.3 billion of government money. Pausing or changing the phasing of projects rarely saves money and almost always increases costs. It soon became clear that these cuts were not enough for the government. Gilligan had been brought in to deliver the final blow and his paper for Policy Exchange gave Sunak the justification for doing it.

In July, with his position increasingly untenable, HS2 chief executive Mark Thurston, who had struggled to keep the programme on track for six years, found himself scapegoated by the government and announced his resignation. Thurston – an effective if sometimes abrasive chief executive – had had enough of dancing around ministers and changing the plan with every new regime. From being the highest paid public official in the UK, he moved on to become chief executive of the privatised water company, Anglian Water. For more than a year (Thurston left in September 2023), the government and the board of HS2 failed to appoint a new chief executive. According to insiders, other senior roles within HS2 were also not filled. The organisation was rudderless. Thompson, the civil servant chair, took on the additional role of interim chief executive.

With Thurston effectively sidelined, Number 10 asked the Treasury to examine the total HS2 costs more closely over that summer of 2023. The plan to cancel the northern leg of the line was carried out in deep secrecy while Harper assured the public that it was definitely going ahead. The DfT was kept out of the loop about the plans as long as possible in case they kicked up a fuss. As Gilligan reasoned at the time, the thing had been their whole lives.

There were inklings that a scaling back was going to occur. A press photographer had caught Jeremy Hunt, the then chancellor, outside Downing Street with a document headed 'chancellor and prime minister bilat' on 12 September 2023. The papers showed a savings table which itemised the cost of each phase. Gleeful national newspapers and the more anxious trade press ran articles about possible cancellation. The High-Speed Rail Group, which lobbied on behalf of the industry, issued a statement claiming more changes would be a disaster for the Midlands and the North. There were heated debates in the House of Commons, with the chair of the transport committee, the Conservative MP Iain Stewart, arguing 'Communities would have been enormously impacted for no great benefit' and urging the government to 'either do it properly or don't do it at all.'

Andy Street couldn't believe that the future of HS2 was on the table and even produced a quote explaining why the Treasury should keep the project under review. He says now that when he asked Number 10 for details, he was told that cancellation was not an option. He believed that Sunak would tell him if anything radical was planned. He was far too trusting. He had based a whole transport and industrial strategy in Birmingham on the arrival of HS2 and its continuation north and found it unimaginable that everything could be swept away with the stroke of a pen. He had seen the second hybrid bill from Birmingham to Crewe go through parliament and the third being prepared and believed that parliament couldn't be overruled unilaterally by the prime minister. But as Johnson had repeatedly demonstrated in years prior, parliament itself isn't particularly powerful when a prime minister decides to ride roughshod over constitutional norms.

As a regional mayor, Street had no power in London. All he could do was place his trust in various prime ministers and chancellors to fulfil their side of the bargain. In such a centralised country as Britain, if you are not in the so-called Westminster bubble, participating in the discussions, it is difficult to influence what is happening. Street lived in a flat in Birmingham, while Gilligan worked in Downing Street, whispering in Sunak's ear that HS2 didn't have popular support in the Midlands.

Greater Manchester's mayor, Andy Burnham, had even less influence on HS2, but he was becoming increasingly anxious. While speculation was whirling in September 2023, Burnham tweeted in frustration: 'The southern half of England gets a modern rail system and the North left with Victorian infrastructure. Levelling up? My a**e.' George Osborne and Michael Heseltine intervened in an article in *The Times* towards the end of September, calling the rumours of cancellation a 'gross act of vandalism'. They continued: 'It would be an act of huge economic self-harm, and be a decision of such short-sightedness, that we urge the prime minister: don't do it.' Their intervention was helpful to Sunak as proof that cutting HS2 would upset the elites. It didn't matter that Osborne and Heseltine were prominent Tory ex-ministers – they were part of the old guard who had campaigned to remain in the EU and were easily dismissed as only interested in building the economy in metropolitan cities.

Sunak and his people in Number 10 kept telling Street that nothing had been decided. He was even taken to breakfast by one of Sunak's team and told everything was going to be okay. Street felt he was being 'patted on the head'. Keeping Street in the dark until the last moment was important for Gilligan and Sunak because Street had the power to sway opinion within the Conservative Party and had proved that he was prepared to throw money at campaigns to retain HS2. At least one big donor to the party had made known he was considering withdrawing funding because of HS2's uncertainty and investors were publicly announcing their concerns too. Jürgen Maier, a former chief executive of Siemens UK, said of the rumours: 'The business community is in total shock and investor confidence is as low as I have ever seen it in my long years of engaging with our government.'

By the time of the Conservative conference in October, which took place ironically in Manchester itself, the cancellation of HS2 was all anyone could talk about and Street became frantic. He gave an impromptu press conference outside the Midland Hotel where Sunak was staying. The prime minister, Street said, was in danger of 'cancelling the future.' This had 'become a debate about Britain's ability to do the tough stuff successfully, as previous generations of Britons certainly did... And of course now it's become a debate about Britain's credibility as a place to invest.' Street's intervention was an extraordinary piece of theatre and unprecedented for a previously loyal mayor at a party conference. Street says today: 'I didn't go to Manchester expecting to take on my prime minister, but I've always known that being mayor was first and foremost about loyalty to your place and so it was not difficult for me to put place before party.'

'ROBBING THE WHITE ELEPHANT'

Andy Burnham was also galvanised into action, and in an unprecedented move for a Labour metro mayor turned up at a conference event run by the High-Speed Rail Group to argue to a room of Conservative Party members that HS2 was vital to Manchester and the country's economy. But it was all too little too late. The day before Sunak's speech to the party conference, he summoned Street to talk to him in a hotel room in Manchester. Street imagined optimistically that Sunak would 'come to an accommodation of some nature'. That wasn't what happened. Street remembers 'being spoken at for a period of time' but not told definitively about the scaling back of HS2.

Street later discovered that Andrew Mitchell, then the MP for Sutton Coldfield – who knew about the HS2 plans – had told the prime minister that Street needed to be 'treated properly.' And so, later that day, Sunak's chief of staff Liam Booth-Smith met to inform him exactly what the speech was going to entail. Allowing Street to believe, until the last moment, that HS2 to the North might still go ahead was an act of extraordinary treachery. Street subsequently learned that Sunak had recorded a video in London before he left for Manchester justifying the cancellation. All the exhortations by both Street and Burnham at the conference couldn't have stopped the cancellation. The decision had been taken days, if not weeks, before.

Today, Street is still a member of the Conservative Party although he narrowly lost his seat as West Midlands mayor to the Labour candidate Richard Parker the following May. Although Street maintains that he fell out irreconcilably with his party over HS2, he was elevated to the House of Lords in December 2024.

During his main party conference speech, Sunak announced that HS2 would no longer go to Manchester or Crewe. It would stop at Handsacre Junction just north of Birmingham to join the West Coast Main Line. He also said that the line would not go to Euston unless private money could be found to fund the station and the approaches, but that he expected 10,000 homes and other businesses to be built there to fund the railway station. He promised that the £36 billion which had been earmarked for HS2 to Manchester would instead be spent on 'local projects': rail services which joined many more cities and towns of the North. The plans included building the Midlands Rail Hub, which would connect fifty stations, developing a Leeds tram system, electrifying train lines in north Wales and extending the £2 bus fare until December 2024. There was also money to fill potholes and improve roads.

None of the projects outlined would have the same national impact without HS2, few of them had detailed plans attached and most of them had been announced before. The Midlands Rail Hub, a series of line upgrades round Birmingham, had been promised periodically since the 1990s. When journalists questioned the Department for Transport about the actual schemes which were to be funded, they were told the list simply contained 'examples'.

Henri Murison, director of the Northern Powerhouse Partnership, who speaks for businesspeople and civic leaders across the region, said that the cancellation would set the North back a hundred years. He uses a fireplace metaphor to demonstrate the interconnection of the schemes. 'One high-speed line for the west, one for the east and – across the fireplace – a lintel, Northern Powerhouse Rail.'

Andy Burnham was furious, speaking of 'frustration and anger.' He told the BBC: 'It always seems that people here where

I live and where I kind of represent can be treated as second-class citizens when it comes to transport.' In other interviews he appeared defiant, claiming that he didn't believe the government really cared about the North any more, which was always being 'palmed off'.

Even players like Andrew Stephenson, the former HS2 minister, considered resigning because he was so outraged by what had happened. He didn't, because, as he says now, almost regretfully, his constituency wasn't affected and he was no longer rail minister.

The cancellation was the death of HS2 as a serious railway line. A high-speed train running from Old Oak Common to Birmingham has very little value and it is unlikely there would be many buyers for the line concession, as there had been for HS1. The government will be forced to run it at a massive loss. The more you look at the ramifications, the more shocking the decision is. Instead of solving problems of northern connectivity, the new high-speed line is now likely to decrease reliability and speed on the West Coast Main Line north of Birmingham and make journey times longer, because there will be more trains running on the most congested part between Birmingham and Preston. The congestion will mean even more freight will transfer to the roads and the M6, already one long traffic jam, will come to a standstill.

Brand new stations originally designed to have capacity for eighteen high-speed trains an hour in each direction will now either be redesigned or scrapped. The station at Old Oak Common may have to accommodate trains turning around, which was never the plan. Airports in Birmingham and Manchester, which would have taken some of the load off

London airports had HS2 connected to them, won't see anything like the same benefit – hence the new government's plan to resurrect a third runway at Heathrow. HS2 trains, if they are only going to run on such a short piece of line, appear redundant, although multi-million-pound contracts have been let to construct them and the future of the train-building industry at Alstom Hitachi relies on the HS2 order.

Euston was even more of a fiasco. The prime minister said that instead of going ahead with the station and approaches, the government would create a new Euston Development Zone which would build thousands of new houses and business opportunities, plus a station to 'develop the capacity we need'. How was not clear. The station would only, he added, have six platforms and would be constructed with private sector funding. For a while it looked like a HS2 terminus at Euston was off the cards. What private consortium could possibly borrow £4.8 billion for a risky station with limited development space around it? That kind of money is only available to governments. When Michael Gove, then the secretary of state, doubled down and suggested there was room for 10,000 homes, everyone laughed.

Sunak also paused the tunnels from Old Oak Common and cancelled the tunnel which would join up Euston with Euston Square underground station – an extra piece of work TfL had quietly snuck into the HS2 budget. A DfT briefing which prompted Sunak's decision showed the cost for Euston and its approaches had ballooned to £7.5 billion.

All work stopped on the HS2 site at Euston after the announcement. Around 1,000 workers were laid off and redeployed and only a handful of security guards remained. The companies hired to build the Euston section of the line were blindsided, although by this time the HS2 project was so toxic and political for all involved that nothing really surprised them anymore.

**

Some 30,000 jobs that HS2 would have sustained in the Midlands and the North for at least the next decade (and the skills and innovation which would have resulted from them) were deleted with a stroke of Sunak's pen. And the cancellation drove up the cost of future construction in the UK dramatically. What business in the world would invest millions in winning a UK government-backed infrastructure contract if the project could be cancelled unilaterally by the prime minister, even after an act of parliament had agreed it?

The fact was, and still is, that the line north of Birmingham would have been much cheaper to build, because the track ran over easier terrain nearer to the current West Coast Main Line. And the line would have brought huge benefits to the North in terms of jobs and investment. Even east–west links between Liverpool and Manchester would have been much faster and formed the foundations of a large investment zone in the North West. Lessons learned from the first phase could have been applied to the second phase, with better civil works contracts and more modular designs.

**

Sunak's scaling back of HS2 was catastrophic, a criminal waste of money and an insult to the north of England and Scotland. The project had been scrutinised for years and years in parliament and yet he was able to fundamentally neuter it, without even consulting the House of Commons. There's another piece of fallout too, which some think, if there is eventually a public inquiry, will be significant. And that is the role of Bernadette Kelly at the Department for Transport. It appears that the Department for Transport and the secretary of state were only told about the cancellation a few weeks before it happened, but many within the civil service believe that Bernadette Kelly, the permanent secretary at the DfT, should have written to the secretary of state and asked for 'ministerial direction' as soon as she found out.

Permanent secretaries are accounting officers for government departments and can ask for ministerial direction if they believe a government decision is beyond the department's legal powers or spending budget, doesn't meet the 'high standards of public conduct', might not be value for money or if there is doubt about the proposal being 'implemented accurately, sustainably or to the intended timetable'. A request for a ministerial direction only has to meet one of these criteria: and you could argue that the cancellation of the western leg could, at a stretch, have met all four.

When a ministerial direction is requested, the secretary of state must respond to give their justification for overriding advice. Several ministerial directions are requested every year. In 2022, for instance, the permanent secretary at the Home Office wrote to Priti Patel over the Rwanda refugee scheme, pointing out that the scheme might not save money. Had Kelly challenged the secretary of state, Sunak and Gilligan's 'secret' decision would have been out in the open and the government would have had to reflect on the consequences in public, or at least justify them.

As it was, Kelly was supine. She allowed the prime minister to announce a decision at a party conference with major financial consequences for both the scheme and the whole country without challenge. Indeed, she went further and issued an official assessment basically backing cancellation. At a Public Accounts Committee hearing in November 2023, just over a month later, Kelly struggled to answer questions about the full business case for curtailing the line (and private investment for Euston), admitting that she had only been working on it for a 'matter of weeks' before the announcement. She told committee members:

> I was clear from the start that this would require an accounting officer assessment, and I gave, it is fair to say, considerable time and attention to that assessment, working with my analysts, finance colleagues, Treasury colleagues and the Treasury officer of accounts. It was a matter of weeks, but I am confident that we did the work thoroughly.

A year later and the accounts for HS2 show that the cost of cancelling the leg to Manchester was an estimated £2.2 billion, including the more than £1 billion preparation work before the bill was approved by parliament and £850 million of work that had just started. In total, £474 million had been spent on compensation. Some £153 million of designs for Euston station (incurred between 2021 and 2023) were also written off, because the government had announced the station's platforms would be cut to six. Taking the total bill for scrapped Euston station designs to more than £250 million.

But there were more costs that didn't show up in HS2's figures: stations, junctions and trains on the London to Birmingham part of the line, all of which might still have to be redesigned and recommissioned to fit what was ultimately a different project. The business case also made no sense if HS2 only went from London to Birmingham and not beyond because few of the benefits to the rest of the network would be realised.

***

Sunak's decision was a triumph for Gilligan and the HS2 sceptics scribbling away in their London Victorian terraces, pretending to care about lines across the north of England they would never have to use. Meanwhile, some in the civil service breathed a sigh of relief. Business as usual could be resumed. The Treasury could return to its comfort zone, declare the high-speed rail experiment a horrible and expensive failure and dole out relatively small sums for transport projects in the North while balefully advising government ministers that realistically, growth only happened by agreeing privately funded extra runways for London airports and 'Silicon Valley' projects in Cambridge and Oxford.

17

Our Friends in the North

In Spain and France, regional leaders vie with each other for high-speed stations, ever alert to how they might increase the wealth of their region. Not only is a stop on a line a mark of prestige, but leaders also recognise the wealth high-speed rail can bring to their area. They petition governments and ministers and beg them to invest.

In Britain, many council leaders in the north of England were a little more lukewarm about the benefits of HS2 and were easily diverted into arguing instead for east–west routes across Britain. This sometimes 'too cool' approach meant it was easier for opponents (particularly in London) to argue that HS2 would only bring benefits to the capital and not to northern cities. The tragedy is that it was only at the last minute that northern leaders understood what was about to be lost, and by that time the detractors had won the argument.

The biggest advocate for HS2 outside London was former mayor of the West Midlands Andy Street in Birmingham. Street is a small, wiry, energetic man in his early sixties when we meet for tea at the Grand Hotel in Birmingham. He's not an ordinary politician, nor a traditional Conservative; he's more in the radical liberal tradition of the nineteenth-century lord mayors of Birmingham, Joseph and Neville Chamberlain. From the beginning he understood and championed the economic benefits of HS2. He's at a loose end now that he is no longer mayor (he lost narrowly to Labour in May 2024), but he's still passionate about the city and incredibly angry that HS2 was cancelled. Although the line will run into a huge, brand-new seven-platform Curzon Street station, which is under construction, the train won't go beyond a further junction at Handsacre just to the north of the city included so trains can continue on the existing rail network.

Street has been a proponent of HS2 ever since he started working in Birmingham, years before he became mayor. As a businessman in the city – he was chair of John Lewis until 2016 and led the local business board (the Greater Birmingham and Solihull Local Enterprise Partnership) – he saw HS2 as 'foundational'. He explains:

> It was clear from the beginning that it was a 'must secure' because it was seen as a critical piece of investment on the back of which we could make the case about economic recovery. So this wasn't something I just jumped on when I became Mayor. I can look back at the economic strategy of the LEP and in the first paragraph, it said 'building on HS2'. It was absolutely a foundational piece and the research that was done by, I think, KPMG. It was really clear that the West Midlands would be the biggest beneficiary of interest. So it makes sense, because we were going to be at the centre of this thing.

He dismisses the idea that all the benefit would accrue to London and that it would make Birmingham a commuter town.

That is such a London-centric view of the world. The thing we needed more than anything else was not another railway line to London. We've got a good connection to London, there's eight trains an hour to London. What we actually needed was better connectivity to the rest of the country – and the most congested part of the network is Birmingham to Manchester and Birmingham to Leeds is just a joke. So, what we wanted was to be at the centre of something that was national, not just to be a little faster to London, and of course the whole capacity argument was key to that as well.

He'd also persuaded business, on the promise of HS2, to relocate to the city: HSBC had transferred its headquarters from Canary Wharf. Goldman Sachs and PwC all have substantial bases there. He argued that Birmingham had the 'best in moving-in stats', encouraging people to move to the city from around the UK, attracted by the promise of connectivity and increasingly good jobs.

'If you were the centre of something,' he argues, 'you would be drawing inward investment and there would be better jobs here so that people didn't need to get the jobs in London and there is limited evidence that this is the case *to some extent* from the first parliamentary approval.'

The centre of Birmingham feels like a modern European city today. It is walkable, with many pedestrianised streets. A comprehensive tram system is being built, not only to connect the city with a link to the HS2 terminus at Curzon Street, but also the region itself with a line out to Wolverhampton. Street also viewed HS2 as rebalancing the economy and tilting it towards the North. The city was to benefit from an enormous new station and to become an interchange to the North, whether to Manchester or Liverpool via Crewe or to Leeds. Unlike at Euston where there had been constant bickering with Camden Council and Network Rail about the station and its approaches, the situation in Birmingham was far more harmonious. As Street reflects:

> We know one of the success factors around the world was the alignment between local, regional government, private sector engagement, and we had that in 2011 and in 2015. In the first papers we were going to have it opened in 2022 in time for the Commonwealth Games. We were in that sort of aligned view of the world.

While Birmingham has at least retained a line into the city, even though the benefits will be far less than if the full line was built out, Manchester and Leeds completely lost out on the housing, jobs and investment which might have helped with developing a northern belt able to compete with the country's London-dependent economy. HS2 was as vital to these cities as it was to Birmingham and the West Midlands. HS2 was, as Street says, 'a national endeavour, a once in the century project.'

Throughout the 2010s, the leaders of these cities weren't nearly as active as Andy Street in fighting for HS2. If they had been, might the outcome have been different? They were hampered by various obstacles, not least that the three hybrid bills – the London to West Midlands phase, the West Midlands to Crewe phase and the Crewe to Manchester phase all went through parliament at different times. The first hybrid bill only covered the line between London and the West Midlands, so discussions over that part of the line were what generated media furore. Disproportionate amounts of parliamentary time were spent on that segment of the route, with thousands of petitions from residents and businesses in London and the home counties. If you weren't concentrating, you might have thought HS2 was only going to be built to Birmingham.

George Osborne had attempted to rectify this with his Northern Powerhouse programme, launched in 2014. The idea was to link the 'core cities' of Hull, Manchester, Liverpool, Leeds, Sheffield and Newcastle and invest in science, technology and manufacturing there, as well as to improve educational attainment and skills. The Northern Powerhouse was part of 'rebalancing the economy' which Gordon Brown had been so keen on in 2009. HS2 was key to these plans, not only because HS2 itself would join the cities but because even HS2 trains on conventional lines would increase connectivity.

For northern Labour leaders, to have Osborne, a Tory chancellor, prancing all over their territory was annoying to say the least, and very possibly an electoral threat, especially a year out from the 2015 election. Until 2017, the North didn't have metro mayors (who were elected as part of the Northern Powerhouse project) so each leader spoke for their much smaller area and the interests of the city of Manchester didn't always align with that of smaller towns around it like Stockport, Bolton or Rochdale, particularly on strategic transport issues. Until the advent of Andy Burnham, no one leader could claim to be the consistent voice of Greater Manchester who would be listened to by government. In West Yorkshire it was worse. Tracy Brabin wasn't elected until 2021.

Even if there was little outright hostility, there wasn't much enthusiastic national Labour support for HS2 either. Ed Balls had wobbled in 2013, Ed Miliband rarely spoke out in its favour (especially as HS2 affected residents in his Doncaster constituency) and after 2015, Jeremy Corbyn and his shadow cabinet were more fixated on nationalising railways than building them. What was even more problematic for the HS2 project, and allowed HS2 opponents to spot a gaping hole in Labour support, was that Corbyn and even some of the northern leaders said they would like east–west routes to be improved *before* HS2, a project they were apt to characterise as investment in the South East and London. In an interview with the *Northwich and Winsford Guardian* before the 2019 election, Corbyn laid out the position clearly:

> We supported the construction of HS2 in the sense of it improving the rail infrastructure of the whole country... Personally, I think far more investment should be going into railways in the north of England rather than London and the south-east, and we would support a Crossrail for the north. In any review that the Government is undertaking we will make our case for the inadequacy of rail investment in the northeast and northwest, and the necessity of a much better trans-Pennine link, electrified.

Others were more blunt. The then mayor of Liverpool, Joe Anderson, a colourful figure who was arrested by the police as part of an anti-corruption operation in 2020 (and charged in 2025), told a Lords economic affairs select committee in 2014 that although he supported HS2 to improve the North's infrastructure, he too would rather have cross-country links (sometimes referred to as HS3). 'What I would also say, my Lord, is that as far as I am concerned, if I had a choice between HS2 or HS3, I would go for HS3 all the time, because it is seriously about connecting cities to drive economic growth in those cities to genuinely rebalance the economy.'

Richard Leese, the leader of Manchester Council until 2021, was passionately in favour of HS2, but he faced opposition from the public and the anti-HS2 lobby among the ranks of the local Labour Party. In the same committee hearing, he was challenged about the lack of public backing for HS2 in 'the north'. A YouGov poll had support at only thirty-two per cent and opposition at forty-seven per cent. The committee of unelected peers were interested to know why he was supportive and the majority of the population opposed. His reply shows a certain defensiveness:

> I think the reason that not just I but the entirety of the Greater Manchester Combined Authority are supportive is that we see the development of High Speed 2 as being essential to the long-term economic future of not just Greater Manchester but the north. Perhaps a hint about some public scepticism is that they will look at now [sic], not twenty years' time. We are talking about a network for the future and, indeed, a network that is intended to give capacity for at least sixty years beyond that. The argument is relatively simple: it is that good transport is absolutely essential to a thriving economy.

When Leese stepped down as leader in 2021, one of the memories shared about his tenure was how 'on one wet Wednesday night in 2014, he brutally demolished the arguments of the anti-HS2 lobby and continued his unrelenting hectoring of the defeated opponent long after the motion was carried.'

Leese understood that HS2 would not only benefit Manchester, but would have a knock-on effect in the smaller surrounding towns, just as the Shinkansen has proved in Japan. But the local population in places like Wigan didn't see that and certainly didn't like the idea of the countryside being torn up for HS2's infamous Golborne Link, which would take HS2 up the West Coast Mainline to Scotland through towns like Preston and Carlisle and have faster links to Birmingham and London. So, in the early stages of HS2 planning, there was political support in the North, but the public wasn't onside and the message was muffled by internal politics and competing priorities between cities and towns. The idea that east–west links would be better for the North was proposed often enough by northern politicians that it muddled the message and played right into the hands of HS2's opponents in London.

Some I've spoken to suggest that northern leaders took HS2 for granted, as you might take it for granted if you were offered a large expensive car from a rich benefactor. Leaders didn't feel ownership over HS2 because they hadn't committed any of their money to the high-speed line. They hadn't had to make a financial contribution (as regional leaders have to in Spain and the mayor of London did for Crossrail). Since the high-speed line was seemingly in the bank, local leaders thought it acceptable to argue with the government over every detail – right up until cancellations started to be announced. These quibbles dominated discussions: where in Sheffield would the station be? What should HS2's Manchester Piccadilly station look like? In 2022, Manchester's politicians and businesses petitioned parliament against HS2 because they were only being offered a surface-level 'turn-back' station and not an underground through line which, they argued, would have improved connectivity from the centre of Manchester.

Manchester's mayor, Andy Burnham, projected a much stronger voice to HS2 but he only came onto the scene in 2017. The year Burnham was elected, HS2 was becalmed. Theresa May was juggling multiple Brexit votes and the Labour Party was recovering from having tried to overthrow its leader Corbyn. HS2 was barely on the political agenda, and certainly not the leg to Manchester. There was little Burnham could do to push it forwards, and he was barely included in discussions. He only intervened when he thought the line was actively under threat, as in 2020 when rumours were flying that Boris Johnson would cancel the Manchester segment. Burnham's frustrations were evident as he told *BBC Breakfast*:

> It's the same old story. London and the South gets whatever it wants, and it's all about penny-pinching in the North. I would say this to the prime minister and the government today: This is your first big test of your commitment to the North of England and we're watching very closely. In my mind there's no justification at all for doing one thing between London and Birmingham, and doing something different in the North. If you're going to do it, do it properly. Don't do it by halves.

But the North was about to be betrayed. Over in Leeds, much of the regeneration of the southern part of the city had been predicated on HS2. The fast line would not only have linked Leeds with Birmingham but would have cut the journey time between Yorkshire's biggest cities Sheffield and Leeds to twenty-five minutes. Such a fast link would have accelerated growth for both, making the cities easily commutable and creating a viable investment zone and profitable area for new affordable housebuilding. Travelling between the two currently takes up to an hour by both train and car.

Leaders in Leeds and Sheffield were supportive of HS2 but wanted other rail projects to go ahead at the same time, for instance the Midlands Engine Rail (to connect up with Nottinghamshire) and Northern Powerhouse Rail for east–west links. Although she worked closely with Burnham, Leeds City Council leader Judith Blake (now elevated to the House of Lords) was still not beyond declaring that the eastern leg would bring double the benefits of the western leg, arguing that that segment should be prioritised. The city bundled its rail demands together in an unholy soup, which made it easy for Whitehall to say: 'Enough now' and cancel everything.

Now, all Leeds has left is the electrification of the Transpennine Route and some money to develop a metro service to Bradford, but not the cash to actually *build* it. The mayor of West Yorkshire, Tracy Brabin, who was only in post for a few months before the cancellation of the eastern leg, said she found the cut 'outrageous' – as was the subsequent collapse of the Integrated Rail Plan which was supposed to compensate for the loss of the HS2 line to Leeds and Sheffield.

She continued: 'So when the integrated rail plan all fell apart… we did go pretty ballistic… how do you expect us to grow our economy when you can't connect us to Birmingham. This is one of the biggest infrastructure projects in the country and actually this is affecting global confidence in us to deliver infrastructure projects.' She confirmed the new Labour government had told her there was no way HS2 would come to Yorkshire in the current climate.

OUR FRIENDS IN THE NORTH

In Manchester, it looked like HS2 had been given a reprieve. In the early 2020s, Andy Burnham and the new CEO of the Northern Powerhouse Partnership Henri Murison, a former Labour politician himself, were for the first time both able to articulate clearly the benefits for Manchester of HS2 *and* Northern Powerhouse Rail, the proposed railway between Liverpool and Hull, emphasising that the two railways relied on each other to work. Even the newish mayor of Liverpool, Steve Rotheram, was a fervent HS2 supporter.

In practical terms, both were intertwined. The first part of Northern Powerhouse Rail for instance was due to run on HS2 tracks between High Legh (near Manchester Airport) and Manchester Piccadilly station in the city centre, effectively halfway to Liverpool. Indeed, Allan Cook, when he was chair of HS2, identified almost fifty miles of HS2 lines into Manchester and Leeds that could have been used for Northern Powerhouse Rail representing more than fifty per cent of the total new lines needed. And the high-speed connection to Manchester airport would, as Burnham explained recently on Radio 4, have obviated the need for a third runway at Heathrow with passengers from all over the north of England (and even London) being able to fly out of Manchester and Birmingham. Birmingham will still have an interchange at Solihull for the city's airport. Travellers will be able to step off a HS2 train there and be whisked by an automated people mover (APM) – one of those driverless metro-type trains found at airports – to catch a flight abroad.

But it was all too little too late and the northern leaders, Burnham, Rotheram and Street, were deliberately excluded from plans to cancel the western leg in 2023.

The cancellation galvanised the mayors into joint action and to argue that the hybrid bill for HS2 from the West Midlands to Crewe, which had been passed through parliament in 2022 could be repurposed for a less expensive line. In September 2024, the new Labour mayor of Birmingham Richard Park and Andy Burnham launched a report by Sir David Higgins proposing a 'high-speed lite' solution. The railway proposed would be more traditional, could be partly paid for by the private sector and would cost forty per cent less than HS2. But it would be a more traditional high-speed line. The track would be ballasted instead of using concrete slabs and deploy narrower British trains running at 300 kilometres per hour rather than 360–400 (i.e. 185 rather than 225–250 mph). More than twenty business leaders and vice-chancellors wrote to the new Labour chancellor Rachel Reeves asking her to build this line to create connectivity, arguing that this would bring the same economic benefits as HS2 – which they estimated would add up to £70 billion annually to the economy of Manchester and the West Midlands and provide a yield of £24 billion in revenue for the government.

The new train line, they argued, would solve the problem of connectivity between Birmingham and Manchester and free up capacity and congestion on the West Coast Main Line and the M6 motorway between Crewe and the North, particularly for freight. The Labour government has so far refused to back it until the completion of HS2 and instead suggested they will continue with electrification of the Transpennine Route. More plans emerged from the High-Speed Rail Group in 2025 for a line at least to Crewe from which trains could continue north to Manchester on the West Coast Main Line.

All these are piecemeal solutions to a pressing immediate problem which will have to be solved: from just south of Crewe right up to Preston, the West Coast Main Line is running too many trains, and as the busiest rail route in Europe is reaching capacity. Without a new passenger line to Crewe, Manchester and even beyond, the North may well be served with fewer trains, or with half-length HS2 trains with fewer seats from the south and the West Midlands than they are today and those trains will become more and more unreliable. Growth will not remain static; it will be stifled.

Northern leaders are now fighting both for connections south and connections across country, but it has taken them a long time to come up with a coherent argument and to work together. But central government with its top-down approach was more culpable. For all the talk of the Northern Powerhouse, ministers and civil servants didn't think it was important to make HS2 a partnership project with northern cities. They thought they could pull all the levers from Whitehall and disempowered northern mayors and council leaders were most often relegated to the sidelines as supplicants. Andy Street, who was the most comfortable working with a Tory government, did all he could to pull together the West Midlands to work with central government – and was repaid for his loyalty with outright betrayal.

As Andy Burnham said on the day Sunak announced the cancellation of the western leg to Manchester: 'They spent billions tunnelling it under fields in the south of the country, but then there's nothing left for us. It goes to the heart of the way this country has always been run.'

18

In the Path of the Ghost Train

The cancellation of HS2 should have been a great win for Deborah Mallender. She is a campaigner who lives near the village of Whitmore which was indirectly on the path of HS2's Phase 2a between Birmingham and Crewe. Whitmore is a beautiful village with houses in the centre dating back to the sixteenth and seventeenth centuries. There's a stream and farmland stretching below it. The place feels ancient and slightly forgotten and there's still a feudal feel about relationships. The area had been the scene of massive protests against HS2 and Mallender was one of their unlikely leaders. I met her in the car park of Whitmore Tea Rooms, which is an old coaching inn dating back to the seventeenth century, where travellers between Stoke (just five miles away) and Market Drayton in Shropshire would stop over. It was Sunday and the tea rooms didn't open until 12 p.m., so we sat in the small outdoor seating area as the staff prepared for the day. Mallender was keen to show me the effect HS2 has had on the land, even though not a single sleeper has been laid.

She pointed down the hill. The proposed high-speed line would have run alongside the current West Coast Main Line, criss-crossing local farms. Some land was bought by HS2 because the line runs right through it, while other areas slightly further away were considered blighted because of the noise and people living in those houses were eligible to apply to have their homes bought by the government. It was a concession to rural areas. As a result, many houses in the area are now empty. When Sunak cancelled the route, he very quickly removed the 'safeguarding' from the land, which, in principle, meant everything the government had bought to build the line could now be sold off. This is *incredibly* unusual. The land for Crossrail 2 in London – which no one expects to be built anytime soon – is still safeguarded. The move was seen by some as an attempt by the government to 'salt the earth', preventing an incoming Labour government from reviving the northern leg of HS2 or any other similar project.

However, the land has since proved unsellable due to the 1,800 boreholes littering the landscape and uncertainty over whether some kind of railway line might be resurrected. These large holes in the ground, sometimes up to a hundred-feet deep, were used to test the geology and monitor the ground water and will need to be filled in, at significant cost. For the meantime, it's a spooky, abandoned place. Large CCTV cameras stare down at you as you wander around country footpaths, of which there are now surprisingly few. Fields have been cordoned off for electricity works, but despite ominous notices forbidding trespassers, no works have taken place.

Mallender is an energetic woman, a researcher and lecturer. She has a law degree and is currently caring for an elderly friend. She first heard of a high-speed rail line in 1996 when she was working for Stoke MP Joan Walley, who had marched into the constituency office and told her a railway line was in the offing. It wasn't HS2, but a plan, long buried, to extend the Channel Tunnel link up north to Manchester and then Glasgow. Ever since, Mallender had been suspicious the land might be used for another line north and HS2 only confirmed her suspicions. Mallender administers the Madeley and Whitmore Villages Stop HS2 group, run through Facebook. Like all the Stop HS2 movement's Facebook pages, it combines local gossip with intelligence about HS2's operations, plus the odd conspiracy theory. Occasional lapses are forgivable given the secrecy with which the company operates.

She's a forceful character, straight out of a Staffordshire *Midsomer Murders*. She wears sensible shoes and drives a red Land Rover Defender. This campaign is her life's work and she has appeared in several newspapers and television reports. Most recently, Austrian radio got in touch about the role of STRABAG, the HS2 contractors, part-owned until recently by Russian oligarch Oleg Deripaska. She drives us up to the 'ghost village of Whitmore', a more modern suburban development above the older village. This part of Whitmore has been called the millionaires' row of Stoke. The road is wooded and dark, the houses are large and were once home to solicitors and accountants. Some are only accessible down long winding drives while others are slightly more modest. It feels like an abandoned American suburb; the lawns are still well kept but the properties have an unloved air. The owners have all sold to HS2 to avoid being near the path of the train.

Cameras are pointed towards the road. Quite a few houses have been rented out at cut-price rates; two became cannabis farms which have now been closed down, another a squat. It's not hard to see why. They're within easy reach of the M6, secluded and managed by a government landlord whose main business is not property. As we approach one mansion, two white security cars with yellow and white striped markings appear from a driveway to greet us. The men in each car stare. It is clear they spotted us on the CCTV and that Mallender is known to them. No words are exchanged, but it suddenly feels as if we have crossed the border into a secret state.

The cars belong to the security firm Control Risks, which signed a £95 million contract to guard HS2 sites in January 2024. Their website describes them as a 'specialist global risk consultancy' which 'helps organisations succeed in a volatile world.' They claim expertise on the risks associated with the outcome of the US election and cybersecurity in Brazil. A far cry from the Whitmore Tea Rooms.

Around fifty houses here have been sold off. Locals who wanted to move say they 'were browbeaten into accepting a lower offer than the market' because of the cloud of HS2 which hung over them for so long. As one councillor said: 'They now feel full of anger every time they drive past their former home and see that it is still empty.' It's hard to say whether this is true. People who sell their property sign confidentiality agreements and are not allowed to talk publicly about how much HS2 paid them and the amounts people expect to raise from a house sale are not always realistic.

HS2 currently has few plans to sell the houses on and even fewer to sell them back to their original occupants, many of whom have moved away. Mallender drives us down the hill. We stop by a substantial area of land, much nearer the proposed line. It's been fenced off and is being used for activities about which the locals know very little. One week, about seventy vehicles were spotted coming in and out, Mallender informs me. A black Mercedes parked at the gate backs out to get a closer look at us as we arrive, and we scarper.

While the suburban professionals lived up the hill, the valley where HS2 was planned is prime farmland and owned by two landowners whose families have been in the area for centuries: the Fitzherberts and the Mainwarings. Ben Fitzherbert, who runs the Swynnerton estates, is the son of Lord Stafford and heir to the barony: the family acquired the land through marriage in the sixteenth century. Edward Cavenagh-Mainwaring, equally aristocratic, runs a large dairy farm. His family have been there for nine hundred years. Each family has a gastropub named after them – the Fitzherbert Arms and the Mainwaring Arms, set within five miles of each other.

Swathes of both men's land was bought up by HS2 for the route between Birmingham and Crewe. Edward Cavenagh-Mainwaring eventually sold 105 hectares (260 acres) of his dairy farm just five days before the route was cancelled. He's a friend of Mallender's and angry about the process, believing still that he didn't receive enough compensation. He is scathing about HS2, not only because of how aggressively they treated him during the process, but also because of the enormous amounts of money wasted on the project. He was particularly aggrieved that HS2 poured herbicide over one of his wildlife meadows and then brought in tons of water from outside to create a newt pond for great crested newts whose habitats were being destroyed elsewhere. It was done, he believes, with an arrogance and total disregard for his local knowledge.

Fitzherbert, on the other hand, trained as a surveyor at Knight Frank in London and has a whole range of interests on his various estates across Staffordshire and Shropshire – from commercial property to agriculture. He currently has the dubious honour of having the most private land bought by HS2. The exact amount bought or the price paid by HS2 is unclear, but the line, before it was cancelled, was set to run in front of his large country house in Swynnerton village. Mallender considers Fitzherbert a man who drives a hard bargain and made a lucrative deal with HS2. Meanwhile, the landlady at the Fitzherbert Arms says he is a regular kind of person who was delighted when the route was cancelled and celebrated with his family and locals in the pub. Fitzherbert now says he will only buy back his land 'at the right price'.

Meanwhile, HS2 was still continuing to take over land they had bought. One purchase was from a specialist fireworks business run by Simon Boote. He was very surprised that HS2 went ahead with the arrangement to reclaim his lease, as it was finalised six months after the stretch of route to Crewe had been cancelled. The area Boote's company occupied included the site of a large semi-sunken nuclear bunker where, during the Cold War, it was envisaged the central government administration of the region would be conducted. Boote used the bunkers to store huge amounts of fireworks he wasn't able to move – and is now denied access to the area.

The people in this part of Staffordshire – unlike the northern mayors – hope that they will now be left alone, the land will be sold back to them, no line will ever be built and they will be able to live and farm in peace again without being interrupted by noisy high-speed trains.

19

A Disaster for Crewe

The arrival of HS2 was to be a lifeline for Crewe. The former railway town was run-down, its shops boarded up and its population struggling. HS2 would breathe new life into the town, bringing jobs, homes and an entire cultural quarter.

The Crewe HS2 Hub Draft Masterplan Vision has an optimistic photo on the front, an aerial photograph of the town with lights mapping out a glittering future. Crewe is at the centre of a joined-up Britain with links to Glasgow and Edinburgh. The boast is unparalleled connectivity: London in fifty-five minutes, Birmingham in twenty-eight and Manchester in twenty-one. Around the station alone, the town hoped to have built 350,000 square metres of commercial floor space, 7,000 new homes and deliver 37,000 jobs by 2043. The document, produced by the local council, is a confection of bureaucratic jargon and optimism, but it speaks to the prosperity Crewe believed HS2 would bring it. Sam Corcoran, the Labour leader of Cheshire East Council, was excited. Suddenly in his town anything was possible. He had the London architects Grimshaw coming up to design the station, the Trafalgar Entertainment Group had poured money into the ageing Lyceum theatre and the large, flattened parking lot in the centre of town was due to have a multiplex cinema.

Crewe was selected as a hub, and plans brought forward in 2016 for it to open in 2027, because it's a hugely important strategic railway junction in the UK, deemed vital if HS2 was to join up with the rest of the railway network. Trains from Crewe go to north and south Wales, via Chester and Shrewsbury; Glasgow via Liverpool, Preston and the Lakes; Manchester and Manchester airport; Stoke and Stafford; and Birmingham and London. The station is also close to other major transport routes, including junctions 16 and 17 of the M6 motorway. In addition, Crewe is home to one of the biggest freight yards in the country at Basford Hall, a mile south of the railway station. The yard, once built for coal, is large enough to handle a lot more freight than it does. The idea for HS2 was that Crewe could be the gateway to the North West. At least a couple of HS2 trains an hour stopping there would mean passengers could carry on to Wales and Liverpool.

But the arrival of HS2 at Crewe also had numerous other advantages. At the moment, Crewe is an old, broken-down junction, belying its strategic importance. The signalling dates from the 1980s and often fails or breaks. Some of the signalling on the independent freight lines is from the Victorian times because transport planners under Margaret Thatcher's government believed that railway freight would be phased out. Lines also cross each other in hugely inconvenient ways. The Cardiff to Manchester train has to traverse four tracks as it travels east to west. The platforms are all in the wrong place. Trains coming up from the south have to navigate some antiquated tunnels and lines, before landing up in the large, horrible snarled-up junction, and the amount of freight which can be transferred onto the railway is limited because the lines are just too congested, meaning more and more diesel lorries slog up and down the M6, to the extent that the motorway is likely to be one long traffic jam by 2030.

HS2 was a miracle cure. For years governments had resisted investment in Crewe because the price tag was too high. Renewing the signalling alone was likely to cost billions. Other expensive renovation works would have caused huge disruption. HS2 allowed train engineers to think big because not only could they construct a new line, but some of the Crewe renovation could be rolled into the project. They could build a new larger station, the 'hub'. The signalling could be digitised (a total of 669 signalling combinations are needed to make it work more efficiently), new platforms added and tracks rationalised. Passengers from London and Birmingham could transfer onto the new HS2 lines, so double the amount of freight could be run up the old West Coast Main Line. Containers could be loaded and unloaded at Basford Hall and then either distributed from there or taken further north. The idea was to run freight trains straight through from Wembley and bring in extra freight trains from Southampton and Felixstowe. The British logistics industry was thrilled! Their leaders started planning massive investment in terminals and warehouses along the route. Hundreds of millions of pounds worth of extra investment might come into the country just by sorting out Crewe and running HS2.

***

Crewe railway station is one of the oldest in the world, completed in 1837, the year of Queen Victoria's ascension to the throne, by the Grand Junction Railway company. It was a stop on the line between Birmingham and Liverpool, and then Birmingham and Manchester. As the railways expanded, Crewe became the significant junction it is today. Early on a factory was established to build locomotives, and the Crewe works, at its height, was the largest in the world, employing thousands of men. The Grand Central Railway company and its successor London and Northwestern Railway (LNWR) designed a town for the thousands of people who flocked to work there. Houses, public baths, doctors' surgeries, mechanics' institutes, churches and schools were all constructed and owned by the company. In 1888, LNWR's chief engineer laid out the magnificent Queen's Park. Crewe was like Wolfsburg was to Volkswagen in Germany; supporters of Crewe Alexandra Football Club are still called the Railwaymen.

As steam engines stopped being manufactured in Britain in the 1960s and England was no longer selling trains around the empire, Crewe stopped being relevant to the national story. The works staggered on, manufacturing diesel engines, but not on anything like the same scale. The grand Victorian station – Grade II listed of course – fell into disrepair. The vast Crewe works finally closed in 1991 and most of the huge halls were flattened and sold off for housing. Families left the town to find employment elsewhere and gradually the shops started to close. There are remnants of the industrial heritage still. The French company Alstom, which took over from the British train company Bombardier, has kept maintenance workshops for the bogies (the chassis which carry the wheel sets) on Pendolino trains, but they only employ thirty-five people. It is planned that new HS2 trains will have their bogies built and maintained there. Bentley Motors is based in the town and Whitby Morrisons workers are still building bespoke ice-cream vans there.

While there is low unemployment in Crewe, the vast majority of the 55,000 population don't have very high skills: one in five simply have elementary qualifications according to the census. The poverty that comes with low wages means appalling ill health. People die early in Crewe. If you're a woman you'll only live to seventy-seven and if you're a man you're likely to be dead by seventy. In leafy Wilmslow East, forty minutes up the road and home to Premier League footballers, the average life expectancy is twelve years longer.

Every other health indicator in the town is horrific too. Cancer rates are high, child development at five is way below average and obesity is through the roof. Walk around and you see empty shops in the 1970s city centre plastered with large sticker-signs offering levelling-up funding for people interested in starting their own business. There isn't much take up. Accessorize and Dorothy Perkins and even Wilko are closed. The Railway Heritage Centre feels down at heel. Much trumpeted in publicity for Crewe, the highlight of the centre is a visit to two magnificent old signal boxes with pre-war signalling equipment, a Bakelite array of hundreds of switches, like a code, impossible to learn. You can also sit in the abandoned APT-P prototype inter-city train and various other locomotives and there's a viewing platform where you can watch the mainline trains.

But it's not a professional museum. The Victorian railway sheds are falling down, the front desk is manned by volunteers and there are a couple of retirees running dusty model train sets which look as if they've been salvaged from someone's garage. The pictures on the displays are peeling. There's some publicity promising a new museum, but the volunteers are gloomy about whether it will make a difference. London architects have already pocketed £45,000 to do a feasibility study, but little seems to have come of it. The connection with HS2 was the opportunity to turn the town's fortunes around and was heavily supported there.

***

More than a year on, ex-council leader Corcoran can't quite believe what happened. The town had so much riding on HS2. Crewe bet the bank and lost everything.

'We never thought it could be cancelled,' he told me. He and his Conservative predecessor had spent £8.6 million the council didn't have. Some £2 million went to Network Rail for station design, £4 million went to Jacobs for technical and transport support and smaller contracts for Deloitte to KPMG. As well as losing some of its own money, the town has also lost investment. There will no longer be a new station as it stands at the end of 2024. The freight companies have shelved their plans. Peveril, the developer who was going to build the cinema in the centre of the town, pulled out, citing both inflation and the cancellation of HS2. Only the theatre remains, and a new publicly funded Cheshire archives building – which will only house half the archives – the other half will be placed in Chester.

It wasn't unreasonable for Corcoran, a cautious accountant, to believe HS2 would happen. The legislation had been given royal assent in February 2021 as part of the High-Speed Rail (West Midlands–Crewe) Act. The legislation, he believed, gave him protection. He'd heard rumours there might be changes in the pipeline in 2023, which became louder in the autumn, but had dismissed them. At worst, he thought, the government might choose to have fewer trains stopping at Crewe. Right up to the last moment, his officers were in touch with the Department for Transport officials about designs for the new Crewe hub. At the Labour Party conference a week before the Tory conference, he met the mayor of Liverpool, Steve Rotheram, and mayor of Manchester, Andy Burnham. They discussed standing firm on HS2 and making sure the promised number of trains stopped at Crewe, but there was no talk about cancelling the entire line.

Corcoran heard the news the next day. For him personally and for Cheshire East Council, the cancellation was catastrophic. The council couldn't recoup any of the money they had spent on consultants. Cheshire East's meagre reserves were all but wiped out and balancing the budget they were preparing that autumn was proving impossible with a special educational needs budget out of control and the rising costs of looking after the elderly locally. Councillors discussed issuing a 114 notice, which is effectively an admission of bankruptcy. Eventually, Corcoran resigned as leader in July 2024 after being unable to hold together the precarious coalition between Independents and Labour.

To add insult to injury, none of the money saved from scrapping the northern leg of HS2 went to Crewe or Cheshire East. The town received little support from the local Conservative MP Kieran Mullan either, who up until then had been a vocal supporter of HS2. He quietly announced he was resigning his seat and was selected for the much safer Sussex seat of Bexhill and Battle, held, ironically, at the time by the retiring rail minister Huw Merriman.

***

It's quite extraordinary that Crewe has been allowed to sink, with all the benefits which HS2 might have brought to the town and the UK railway network, and that Rishi Sunak cared so little about transport that he didn't seem to understand the vast expense and many consequences of cancelling this part of the line north of the West Midlands; that there was no public discussion of a catastrophic decision. Crewe was just seen as an incidental casualty. And Crewe is just the most egregious example along the route. Other smaller places – Toton for instance, where once there was due to be an East Midlands hub station – had planned a major regeneration plan which would have brought housing and growth to the wider region.

There are two points here. First, to have encouraged a struggling northern town to spend so much money on a government project – and then to abandon the people of the town to fend for themselves is callous. Crewe hasn't been compensated and seems to be viewed simply as collateral damage. It is possible HS2 would not have brought the expected economic benefits to the town, that they would have taken years to be felt or that people would have used the high-speed line to escape Crewe and work elsewhere. But with a proper industrial strategy, backed by central government, Crewe could have been successfully revived, especially if there had been better education and super-fast broadband – in 2021 Uswitch identified Wistaston Road in Crewe as being the slowest street for broadband in the country. Learning lessons from other countries around the world, Crewe could also have become a tourist destination with a beautiful railway museum, a smaller rival to York. With good transport networks, decent broadband and a thriving cultural centre, it could have become a great place for people to live in the North West.

From a railway and transport perspective, cancelling the line to Crewe was even madder, borne out by the anger of the rail industry about the plans to stop the line at Handsacre, a junction just north of Birmingham. The line to Crewe was relatively easy and cheap to build compared to other stretches of the line. There wasn't any tunnelling required and the land had been acquired already. Sorting out Crewe solved lots of problems on the West Coast Main Line which, unless a plan is agreed, will continue to be a horrible bottleneck, restricting growth in the North.

So important is Crewe that the rail industry body the High-Speed Rail Group (consisting of the main HS2 players, Hitachi, Alstom, Siemens and Avanti West Coast) wrote a report in autumn 2024 arguing that running HS2 from Euston to Crewe could save the Treasury £3.5 billion on costs because that part of the line was so valuable. High-speed rail consultants Greengauge 21 believed that were the line to be franchised out for thirty years (as the Channel Tunnel link HS1 currently is), the concession would be worth up to £10 billion rather than the £2.75 to £3.5 billion without it. Even if the government didn't sell the concession as part of its nationalising railways plan, the Treasury would still be likely to make a lot more money both on freight and passenger services if the junction and signalling at Crewe were sorted out.

HS2 trains themselves will now be shunted out to a place called Handsacre, currently a building site thirty-six miles north of Birmingham near Lichfield where it will connect onto the mainline. Handsacre will have a slow slip line linking it and that is where the traffic jam will start. Any speed gained on the way to Birmingham will likely be completely lost as HS2 will wait to join the old line and compete with all the other traffic on it, trundling up through antiquated lines and tunnels. Handsacre Junction is quite small, because when the decision to press ahead and redevelop Crewe more quickly was taken, Handsacre, which would have originally been a large HS2 junction, was downgraded. Now that HS2 trains will pass through it, engineers may have to redesign the junction a third time to cope with a stream of HS2 trains and if they do, they will have to buy more land and acquire separate planning permission because changes were not covered by the act of parliament.

As Andrew McNaughton explained to the key transport select committee in November 2023: 'If you are not running most of the trains direct to Crewe, people will have to go back, scrap the [Handsacre] design and change it in full flight. That will cost an inordinately—'

'Inordinately' means £500 million to redesign the junction, and the money saved from scrapping the Birmingham to Crewe line will be even less. With the extra HS2 trains using the line, there is likely to be room for fewer freight trains with fewer goods going up and down the west of the country, less room for more economic growth north of Birmingham and in Scotland and slower passenger trains altogether. Not to mention the sticky questions of train platforms, seats in trains and Manchester.

It is hard to imagine the current government won't find a solution. Even if it's not called HS2, there will have to be a new line going from Birmingham to Manchester via Crewe eventually.

Speeds may be a bit slower, the specifications of the trains lowered, but it is unimaginable that we continue to run so many trains up an antiquated Victorian track past a major failing junction through ancient tunnels. The longer the government waits to come up with a solution, the more money they will have to spend eventually to solve the problem. For now, they have to work out how they are going to finance and finish the first part of the line.

Conclusion

Parliament's Public Accounts Committee's 2025 report into HS2 was scathing.

> The High Speed Two (HS2) programme has become a casebook example of how not to run a major project. It is unacceptable that over a decade into the programme we still do not know what it will cost, what the final scope will be, when it will finally be completed or what benefits it will deliver.

Some eighteen months on from Sunak's cancellation of the western leg of HS2, few are prepared to defend the remains of HS2. But scrapping the whole thing would also have been a nonsense, not least the cost to jobs as around 30,000 people were working on HS2 at the time. So, when Labour came into office in 2024 in a landslide election, ministers were left with no option but to continue.

A lot of engineering structure had already been put in place: a two-mile low-slung concrete viaduct for instance was emerging over the Colne Valley near the M25, spanning water and land. The largest railway bridge in the country when it's up and running, it won the engineering category of the Building Beauty Awards in November that year. Miles of tunnel had been bored under the Chilterns and west London, land had been cleared and foundations for stations dug at Old Oak Common, Solihull and most importantly, Curzon Street in Birmingham. And sections of the thirteen-viaduct Delta Junction in Warwickshire were being cantilevered into place. When completed, this junction will be the most spectacular in the country, crossing rivers, railway lines and the M6/M42 motorway interchange at the gates of Birmingham. Even this severely truncated version of HS2 is having a dramatic effect on the landscape of Britain.

Because little expense was spared, HS2 has also driven new engineering and railway technology in the kind of leap not seen in Britain since the nineteenth century: new methods for low carbon construction, systems to monitor the provenance of materials, robots in tunnelling, AI to help engineers monitor the line's structure for performance changes and uncrewed aerial drones to inspect conditions and assets on the old Victorian line. But HS2 also fell into the trap of sorting out lots of problems for utility companies, Network Rail and local councils (like filling potholes), which weren't directly within the remit of building a high-speed railway.

Back in London, Euston was still in limbo and a building site. After some hemming and hawing, the new Labour rail minister Lord Hendy (former chair of the Euston Partnership and Network Rail) decided the station in central London would go ahead and green-lit the tunnel boring from Old Oak Common, although by May 2025, the tunnel boring machines had not set off. Another alumna of Crossrail, the former CEO, Mark Wild, credited with completing the line, was brought in to sort out the unholy mess.

CONCLUSION

What horrified the Public Accounts Committee members most was that neither Wild nor the department could agree on how much HS2 would cost and committee members had to come up with a figure from their estimates – £80 billion in today's money, now expected to be more like £100 billion. Permanent secretary Bernadette Kelly also wrote to them later to say the full business case would not be presented until 2026.

There was much for Wild to do. He has vowed to renegotiate the contracts to achieve better value for money and produce a 'minimal viable railway', but probably not until the mid-2030s. In March 2025, Euston was still a total mess, with approaches being dug and built for a much bigger and busier service than was now planned. No agreed design for the station existed, nor were decisions taken about how it would be integrated with the current station, nor were any plans made about expansion if more services were run from the North. Camden Council, eventually given some £29 million by HS2 to rehouse and compensate more residents on the Regent's Park Estate, helpfully produced a document saying the regeneration of Euston station and its environs could contribute £41 billion to the economy by 2053 with housing, shops, offices and a life sciences and tech industry. But who might lead a new 'development corporation' to make this happen – Camden or the mayor of London – was still unclear.

Insiders, keen for a silver lining, say there is at least political alignment, with all the main players having worked together previously on Crossrail, including the new transport secretary, Heidi Alexander, who was the deputy mayor of London. Euston is also bang in the middle of the prime minister's constituency. Nasrine Djemai, who was a child when HS2 was first conceived, is now a Labour councillor in Camden, a cabinet member for New Homes and Community Investment.

So what *could* go wrong? A lot, according to the same Public Accounts Committee in 2025, who remained sceptical that private investment could fund Euston and warned that Euston families will be living on a building site for years to come. Meanwhile in Birmingham, Curzon Street station was going ahead despite being far too big for the shortened line: changing the design would have cost too much. And at Old Oak Common, while no one was paying attention, an enormous transport hub (set to be one of the largest in the UK) was springing up, an interchange between Great Western Railway, Crossrail [the Elizabeth line], Heathrow Express and HS2. There is now a real danger that London will once again be where the government concentrates its energy and the real aim of HS2 – to build a high-speed line to join up the Midlands and the North – will be forgotten.

Meanwhile, there are fifty-four 200-metre trains being designed and built by Hitachi Alstom in Derby and Newton Aycliffe, scheduled to run at 225 miles per hour (360 km per hour) up to Birmingham. That too is a muddle. Beyond Birmingham, these trains may well provide a slower service than the current Pendolino because they don't tilt around corners. Each will transport around fifty fewer passengers (550 compared to 607 in the longer Pendolino). The original plan was to couple trains together, but platforms at Manchester Piccadilly aren't long enough for 400-metre trains. HS2 insiders say executives are re-looking at the length of the trains, which haven't yet been built, and may still try to change the design again to make them 250 metres long. Also, a fleet of fifty-four trains is a lot of hardware for a short route – Avanti, for instance, own fifty-six

CONCLUSION

Pendolino trains to run their service from London to Glasgow. But the DfT have already spent the money on the contract and the train-building industry in the UK is relying on the order. HS2 promise the trains – which are using Shinkansen and European high-speed train technology – will be the 'fastest, quietest, most energy efficient high-speed trains operating anywhere in the world'.

In another part of the government forest, the Infrastructure and Projects Authority based in the Cabinet Office still has HS2 marked down as 'unachievable'. Not surprising if you consider the uncertainty over cost and the constant changes of plan. The government has refused to commit to any of the plans put forward by the metro mayors and High-Speed Rail Group to build a new line beyond Birmingham. There will be calls for another body to finish the HS2 project if HS2 Ltd proves incapable of completing what the company has started.

There are many, many lessons to be learned for future infrastructure projects, not just rail projects. Here are those I think are important, based on the story of HS2:

1. Build a rationale for your project that has national buy-in. Building a fast luxury train to Birmingham is not a rationale and increasing capacity on the West Coast Main Line is not a cause to make anyone's heart sing. (Why build an over-specified high-speed train if all you want is a bypass?) But making the Midlands and the North richer with faster connections between London and nearby cities is a cause lots of people could have rallied behind. So is positioning HS2 as a key component of industrial renewal after the financial crash, creating tens of thousands of jobs in the engineering and

supporting industries and building the skills of hundreds of young people and increasing housing.

2. Once a government has a firm proposition, civil servants and politicians should go out and listen to local politicians, businesses, developers and ordinary people in the places that might benefit (and even those which might not) to understand what their interests are, where they think the route should be and what problems might arise locally. Had cities and regions along the route been contributing money and sponsoring HS2, local leaders would have been much more inclined to make HS2 work and drive it forwards. Had civil servants and HS2 understood their stakeholders before deciding the route and the project, more serious thought might have been given to starting the line at Birmingham or Manchester, even if it was politically impossible not to start it in London. Andy Street was onside, had local businesses lined up and saw HS2 bringing his city and surrounding cities and towns more wealth.

3. Don't rush a complex infrastructure project. The route of HS2 was decided, speeds were set at 250 miles per hour and the design agreed by a small group of people in a room in the Department for Transport and then the plan was presented to the public and parliament as a fait accompli only weeks before Labour lost the 2010 election. What engineers dreamt up was more complex than any other high-speed rail project in the world. That's not to say I'm blaming the engineers. Put them alone in a room and tell them to design a world-beating railway and they will! But the DfT, HS2 and the new

CONCLUSION

coalition government should have insisted on a much simpler concept. Instead, through ignorance, fear or hubris, transport secretary Justine Greening suggested more tunnels and chancellor George Osborne asked for a station at Euston that matched West Kowloon in Hong Kong.

4. Make sure that you have expert engineering advice from people in countries who have actually built high-speed rail (or whatever infrastructure you are building). Spain has a highly developed high-speed rail industry and builds relatively cheap lines. Why weren't an army of Spanish engineers and rail experts involved in developing HS2 from the beginning? Couldn't we have bought a HS2 project more or less off the shelf, including new trains? A HS2 insider told me: 'There have been a lot more conversations going on with the Spanish in the last two to three years'. Why not before?

5. Set a realistic budget, or in HS2's case, don't set a fixed budget until you really understand the costs. HS2's original budget was a finger in the air based on cheaper, more developed European high-speed rail projects. The budget didn't take into account political interference, miles of extra tunnelling, extra environmental mitigation and the complexities of bringing eighteen trains an hour on bespoke lines into the centre of densely populated cities.

6. Know your limitations. Many at the senior level in the civil service, HS2 and the government (with honourable exceptions) were out of their depth, not quite grasping the enormity of what they were embarking on, nor understanding

the technical, legal and stakeholder difficulties they would face. Worse, they were afraid to ask for help. HS2 was far more complicated than Crossrail or the Olympics! Many approached HS2 with the gung-ho attitude of real estate developers or the caution of accountants. If government is to build more complex infrastructure, the civil service needs to employ teams at its most senior levels who have the skills to commission and oversee mega-engineering projects and build stakeholder involvement. Senior leaders in the civil service need to stop believing 'learning on the job' is acceptable – and while costs must be understood and controlled, deploying hundreds of young Treasury officials to crawl regularly over the budgets of a mega-project they didn't understand was not helpful and only added to the costs.

7. Be flexible early on. Reduce the speed a bit if local people are worried about noise. Build the line above ground as much as possible. Avoid viaducts and tunnels. Don't plough through areas of outstanding national beauty.

8. Go modular! Forget iconic stations and bespoke bridges. They were for railway magnates in the nineteenth century who wanted to impress investors. HS2 was a fully government-funded project in an age of austerity. Most parts of an infrastructure project should be modular and preferably made offsite like Lego because that is cheaper! Why in a country that loves Ikea do we think all our infrastructure has to be bespoke?

CONCLUSION

9. Be nice. Don't antagonise the people you need onside. Building relationships is as important as building major junctions. All successful high-speed rail projects in democracies are loved by their local communities who are proud to have them – think of the children's Shinkansen bento boxes in Japan and the Spanish mayors who beg to have a station en route. Be generous and firm with compensation, take time to negotiate and think about the effect on the least well-off, not just the most powerful. It matters because values are set early on. HS2 looked like a train for the rich, not least because it treated the richest and most powerful people along the route far better than the poorest and lost sight of the fact that it needed to be a national railway for all.

10. Collaborate. Mega-projects demand it. HS2, the DfT and the government ended up fighting each other and everyone they came into contact with, from local communities to their own agencies and construction companies. This caused rising costs because in the end, HS2 either settled through the courts or paid opponents off. There was not enough careful negotiation. An adversarial and litigious approach to infrastructure building doesn't work. A little less 'masculine energy' might have been beneficial at all levels.

11. Start small. Preferably link the project to a national event or mood. Spain built its first high-speed line from Madrid to Seville to take people to Expo '92. Britain built HS1 not only because ministers were ashamed of our slow line from the Channel Tunnel to London, but because HS1 helped us win

the 2012 London Olympic Games bid. For what national occasion or symbolic moment might we have built HS2?

12. Decide what it is you are building. Is it a completely new line like the Shinkansen with its own everything, or is it a high-speed line like the TGV you want to integrate into the wider network? Be clear from the beginning and design your trains and network accordingly. Any lack of clarity will cause a headache down the line. If France and Japan could nail this down, why couldn't we in England?

13. Have a detailed design before authorising building (think slow, act fast), especially if four different consortia are building four different sections of a railway line. Understand the ground conditions, where the pylons are, what the obstacles are and the details of the build. This was easier said than done in HS2's case because of intense political pressure to proceed as fast as possible.

14. Changes and pauses to a mega-project once major contractors have been employed costs money and can add billions to the final bill. Consider that in March 2025, HS2 was spending £25 million a day on the railway. Even after HS2 had been agreed by parliament and enabling works contracts let, the project was still subject to government review. And ministers were inclined to treat HS2 rather like a Minecraft game, where they could move parts at will. It's extraordinary that Secretaries of State for Transport thought they could authorise extra tunnelling, demand tree-felling stopped, decide to scrap a junction and pause works at Euston station. Not to

CONCLUSION

mention prime minister Rishi Sunak's decision to cancel the northern leg without reference to parliament, which had partly approved it already.

15. Leadership is vital. Have a guiding mind on a mega-project and clear lines of responsibility. HS2 had neither. Even today, it is still unclear whether HS2 or the Department for Transport are in charge of the project and what role government ministers should play. Chairs and chief executives on HS2 have come and gone and hardly ever overlapped, Secretaries of State ditto, DfT officials too. For a crucial fourteen months between September 2023 and December 2024, HS2 didn't have a chief executive and the chair was holding the reins. No one knew what was going on, nor why certain decisions had been made. Crossrail benefitted from the oversight of an experienced team at TfL and the various mayors of London who financially and politically sponsored the project.

16. The private sector cannot raise billions of pounds and construction firms can't carry the risk on contracts which are worth more than the value of their companies. The Exchequer has to bear most of the financial risk for complex infrastructure. Even when the Treasury tries to circumvent paying, recent history shows they mostly end up footing the bill. The government need to be honest about this.

17. The benefit–cost ratio used by the Treasury should be changed. Projects like HS2, designed, as Allan Cook says, to fundamentally change the economy of a country, need different measures

of success. Sometimes, a project like HS2 is needed for the national good and for industrial renewal and benefits and costs should be calculated accordingly.

18. Scrutiny is good – but if scrutiny doesn't improve a project, then what is the point? The number of reports produced by various agencies and parliamentary committees about the finances of HS2 was exhausting. And for what? Nothing changed. The only outcome was growing public resentment and increasingly toxic behaviour by all the parties involved. The concentration on price and budget drowned out any sensible discussion about what HS2 was for and why it was important for the country.

19. Democratic accountability is good, but the hybrid bill committee system for HS2 which heard thousands of petitions from people all along the line was a mixed blessing. A process whereby parliament gives outline planning permission that can't be overturned by the courts is sensible, but allowing thousands of people to petition bored MPs who don't know the local areas concerned is not. Why were parliamentarians allowed to make at least £1.2 billion worth of minor and not so minor changes to HS2 without challenge?

20. Local democracy is good, but the number of permissions (more than 8,200) that HS2 had to apply for along the route was ridiculous. Councils like Buckinghamshire can't be allowed to hold up design permissions for years just because they politically oppose HS2. Government departments like Defra should be encouraging their agencies

CONCLUSION

(Natural England and the Environment Agency) to cooperate rather than litigate or insist on bat tunnels costing more than £100 million. But do we want HS2 and central government to decide haul routes, lorry movements and temporary roads without referring to local democratically elected councillors? I'm not so sure.

21. Protecting wildlife is good but at what cost? The UK is now in the bottom ten per cent of nations globally for biodiversity and we have signed numerous international treaties. In 2024, butterflies declined dramatically. Hedgehogs, water voles and the song thrush are all losing their habitats and disappearing. This is not because of HS2 but over-farming, over-fishing and climate change. Chancellor Rachel Reeves declared that 'we have gone too far' protecting bats and newts, but with HS2 constant ministerial interference was a far greater contributor to rising HS2 costs than environmental mitigations.

22. Lessons learned are important. The government embarked on HS2 without studying why HS1 worked or learning any of the lessons from that project. Nor were many lessons learned from Crossrail applied to HS2. An honest lessons learned document from HS2 would help the government and ministers plan better in the future. What is most striking to me is the extent of government interference and the unwillingness of politicians to be accountable for the rising costs, the constant, often destructive scrutiny of and concentration on the money rather than the benefits HS2 might bring, and the top-down, adversarial culture where there were few attempts to enlist local help in any of the cities along

the way or to formally link HS2 to a national industrial, housing or even transport infrastructure strategy.

Finally, HS2 is not the only mega-project in the world to have caused so many headaches and cost so much more than was budgeted. Oxford Professor Bent Flyvbjerg is a world expert. He reckons that only 8.5 per cent are completed on budget and on schedule and he has a database to support this. HS2 is a textbook example of what can go wrong.

But HS2 is also still the only infrastructure project under way in Britain which has the potential to be transformative for the economy, creating jobs, kickstarting house-building, bringing more prosperity to the north of England and providing a balance to London. A new modular high-speed line should be built out to Crewe and Manchester with extra pieces of track to join Liverpool with Manchester. Ministers need to adopt a totally different collaborative approach involving communities, local politicians, businesses and investors. The 2024 Labour government has a chance to prove that Britain does now understand how to build infrastructure differently and can create a high-speed rail project for the people.

Acknowledgements

I'd like to thank all the people who spoke to me, often at length, for the book and who helped me understand their perspective on HS2. In particular: Lord Patrick McLoughlin, Allan Cook, Andrew Stephenson, Professor Andrew McNaughton, Henri Murison, Jim Steer, Councillor Sam Corcoran, Julian Glover, Richard Lyall, Patrick Diamond, Chris Rumfitt, Lord Mike Katz, Councillor Nasrine Djemai, Deborah Mallender, Emma Crane, David Joyce, Tracy Brabin, Jake Sumner, Lord Nicholas Macpherson, Robert Latham, Lord Andy Street, Baroness Margaret Hodge, Professor Linda Tjia Yin-nor, Karim Palant, Robert Latham, Chris Triffitt, Graham Dellow, Sarah Flannery and Katy O'Donoghue, as well as many others who spoke to me off the record.

I'd also like to thank my agent Charlie Viney, my editor at Oneworld, Cecilia Stein, and her assistant, Hannah Haseloff, who suggested elegant edits to *Off the Rails* and improved it enormously. Many thanks also to Oneworld's publicity director, Kate Appleton, to copy-editor Tom Feltham, and to head of production Paul Nash.

My friends have been very supportive. In particular, Helen Greaves, who always had faith I could write a book like this, Susie Gilbert who took me to Staffordshire and Crewe, Caroline Seymour who came on adventures in Buckinghamshire, and Hannah Robinson who offered help and confidence through the last stages. Final thanks goes to Kate, who always listened, and to my family.

Glossary

AVE
The *Alta Velocidad Española* is the name of Spain's high-speed service run by Renfe, the Spanish state rail company. The first AVE train service ran from Madrid to Seville in 1992.

Ballasted Track
Track built on a layer of crushed stone or gravel which is also placed around the track. Tracks have been built on ballast since Victorian times.

Bogies
Bogies are the chassis which carry the wheel sets.

Crossrail 1
Otherwise known as the Elizabeth line, this urban railway runs east–west across London. The longest part of the line stretches from Reading to Shenfield with branches to Heathrow and Abbey Wood. While the central portion of the track is new, the train also runs on old lines and only ten of the forty-five stations are brand

new (others have been upgraded). It was completed in two stages. The first part opened in May 2022, the second in November 2022, with the full service of twenty-four trains an hour operational a year later in 2023. Transport for London gave the first operator contract to MTR (the company which runs the Hong Kong metro). From May 2025, the concession has been awarded to a consortium of the Go-Ahead Group, Tokyo Metro and the Sumitomo Corporation.

Crossrail 2

On ice since 2020, Crossrail 2 might yet be built – the newly consented British Library extension will leave space for a station. Crossrail 2 is similar to the Elizabeth line – a fast urban train – but would run north–south instead. The core route would run between Clapham Common and Seven Sisters, with a branch via Tottenham Hale, but the train would also run on regional branches down to Shepperton in Surrey and Broxbourne in Hertfordshire. Crossrail 2 is important for HS2 because of the planned Euston stop, which should intersect with the HS2 station.

Docklands Light Railway (DLR)

An automated light metro system first introduced in the 1990s. Over the following decades the DLR was expanded and is now a 24-mile network serving Canary Wharf, the Docklands area and the City of London. The line also extends through Greenwich to Lewisham. The DLR also has a station at Stratford which links it to HS1.

HS1

The high-speed railway from the Channel Tunnel to St Pancras station. The railway was originally called the Channel Tunnel

GLOSSARY

Rail Link but changed its name to HS1 in 2007. Currently, Eurostar run regular trains to Lille, Brussels and Paris on the line. Javelin high-speed trains also run on HS1 at a maximum speed of 225 kilometres per hour (144 mph), stopping at local destinations (Stratford International, Ebbsfleet and Ashford International) on the way to Folkestone. The Javelin also branches off onto conventional lines to reach the Medway towns and Canterbury. The HS1 line is under-used and there are plans for other companies to run trains on the line to the Continent.

HS2 Ltd
The wholly government-owned company which is building HS2.

HS3 and Northern Powerhouse Rail
Northern Powerhouse Rail (NPR) and HS3 are basically the same thing. NPR is officially described as a 'major strategic rail programme', but it's currently unfunded. The idea of NPR is to speed up rail journeys across the major cities of the north of England from Liverpool to Hull via Manchester, Bradford and Leeds, combining high-speed and conventional lines. The new network would aim to free up space on existing stopping lines and allow towns as well as cities to have better, more reliable connections. If HS2 had been built out, fifty miles of HS2 track could have been used for NPR. Now, NPR is in limbo. So far, the government has only committed money to upgrading and electrifying the Transpennine Route, which is only peripheral to NPR.

ICE
The Intercity Express train is a high-speed service introduced in Germany in 1991. The ICE travels at speeds of 190 miles per

hour in Germany and 200 miles per hour once the trains cross over to France. The ICE is marketed mostly to business travellers as an alternative to flights, but like all trains in Germany, the ICE trains run by the state-owned Deutsche Bahn have become increasingly unreliable, plagued by delays and cancellations. The problems have been put down to old infrastructure which hasn't been maintained and the fact that the track and trains are run by the same state company.

Ineco
The state-owned Spanish engineering company which builds high-speed rail and comes under the Ministry of Transport, Mobility and Urban agenda. The company is now helping to build high-speed networks in Morocco and is entering a consortium to build the proposed Australian high-speed line between Sydney and Newcastle.

Jubilee Line Extension (JLE)
The extension of the Jubilee Line east of Green Park to Stratford, which opened in 1999.

Mini Shinkansen
The concept of widening the gauge of Japan's narrow-gauge tracks to standard-gauge tracks so they can take high-speed trains. Trains run more slowly on mini Shinkansen tracks at 80 miles per hour compared to 199 miles per hour on the Shinkansen track, but the mixing and matching has helped form larger networks.

GLOSSARY

Rolling Stock
A general term for all vehicles used on a railway, locomotives, carriages and wagons.

Shinkansen
The official name of Japan's bullet train. In Japanese it means 'new trunk line' or 'new mainline' and is used both to describe the trains and the lines the trains run on.

Slab Track
This is track laid on concrete slabs. The tracks are lower maintenance than those built on ballast, more resilient to climate change and can take heavier traffic, but the concrete slabs need foundations and slab track can be almost twice as expensive to lay.

SNCF
Founded in 1938, the *Société nationale des chemins de fer français* runs all the trains in France and Monaco including the TGV.

TGV
The original European high-speed train – *train à grande vitesse* – was first launched by President François Mitterrand in 1981 and now runs at speeds of 200 miles per hour through France, with services extending to neighbouring countries. The TGV runs on high-speed and conventional tracks. For instance, the approach into Paris and Lyon is on conventional tracks so that the SNCF never had to dig up either city to lay a high-speed line, cutting costs and complexity.

Further Reading, Viewing and Listening

For the history of the railways in Britain, Christian Wolmar is the expert. He's written a lot of learned books on the subject including *Blood, Iron & Gold: How the Railways Transformed the World* and *Fire and Steam: How the Railways Transformed Britain*. Wolmar was not a fan of HS2 and laid out his objections in a comprehensive article for the *London Review of Books*, 'What's the point of HS2?' in April 2014. Although I disagree with him, the piece is worth reading. He also has a podcast *Calling All Stations* and his conversation with Jim Steer, a high-speed rail advocate, is elucidating. You can listen to it here: https://cogitamus.co.uk/calling-all-stations/calling-all-stations-with-christian-wolmar-episode-13-season-2 (all links correct at time of publication).

Terry Coleman's *The Railway Navvies: A History of the Men Who Made the Railways* is an in-depth look at the people who risked their lives – and sometimes lost them – building Britain's railway system. The book is a reminder that the tracks and tunnels used today were dug out by hand in the nineteenth century. *Guardian* columnist Simon Jenkins (also an influential HS2 opponent) has written a beautiful photo book: *Britain's 100 Best Railway Stations*, which goes part of the way to explain why people like George Osborne and Lord Adonis thought building iconic stations was the way to go. Likewise, *Architectural Review* profiled the West Kowloon station Osborne was so taken with: https://www.architectural-review.com/buildings/hong-kong-west-kowloon-station-by-andrew-bromberg-at-aedas.

There are a lot of contemporary literary accounts of railway building. Charles Dickens' *Dombey and Son* looks at how railways affected cities for better and worse. George Eliot's *Middlemarch*, set just before the passing of the 1832 Great Reform Act which broadened the franchise and redistributed parliamentary seats to industrial towns, gives a contemporary view of how railways shaped the politics of the day. I was intrigued by artists' scepticism, which mirrors concern today – William Wordsworth's 1844 poem 'On the Projected Kendal and Windermere Railway' might have been written by anti-HS2 campaigners! Painted the same year, J. M. W. Turner's *Rain, Steam and Speed – The Great Western Railway*, which hangs at the National Gallery in London, also reflects on the tension between man and nature. Art historian Dr Christina Bradstreet gave an illuminating talk about the painting (https://www.youtube.com/watch?v=N8mf9y6ziXA) in which she made a direct comparison between the anxiety over rail in the nineteenth century and the worries about HS2 today.

Samuel Smiles' 1862 *Lives of the Engineers*, which can be read and downloaded for free via Project Gutenberg, is fantastic for its moral portrayal of George and Robert Stephenson. To be read with a *large* pinch of salt, it offers a compelling insight into the status of engineers in Victorian England – a status they sadly no longer enjoy.

FURTHER READING, VIEWING AND LISTENING

Finally, I found Professor Simon Gunn's report for the government, *The History of Transport Systems in the UK* (https://assets.publishing.service.gov.uk/media/5c07d08240f0b670656346e3/Historyoftransport.pdf) fascinating, and one of the very few documents I've read which looks at twentieth-century transport in the round. Published in December 2018, the report is accessible and blissfully short at twenty-seven pages.

The London to West Midlands hybrid bill was finally given royal assent in 2017. There is no one place where all the information about the bill, the debates and the petitioning can be found, a point highlighted by the Hansard Society. Their blog post, 'HS2 Fiasco: What does it mean for Parliament?' (https://www.hansardsociety.org.uk/blog/hs2-fiasco-what-does-it-mean-for-parliament#is-parliaments-approach-to-infrastructure-projects-part) accuses Rishi Sunak's government of having ridden roughshod over both Houses.

It is impossible to track the hybrid bill through the pages of parliament's website (in contrast to other legislation) but the minutes of at least some of the hearings are searchable. The first set of almost 2,000 petitions from 2014 are riveting because they present a slice of life, but they also demonstrate the folly of the petitioning process for such a controversial project as HS2 which would affect so many people. You can read them here: https://publications.parliament.uk/pa/cmhs2/petitions/petcontents.htm.

There are also the pages and pages of supplementary evidence provided by petitioners when they appeared before the committee, including many PowerPoints and maps laboriously put together (https://publications.parliament.uk/pa/cmhs2/cmhs2rev.htm). Looking at a few is enough to make the head spin, but these presentations were important for individuals, worried about how their houses and lives might be affected.

The House of Commons library, in its 2024 report about HS2 for parliamentarians, gives some of the key timelines and events, which is helpful: https://researchbriefings.files.parliament.uk/documents/CBP-9313/CBP-9313.pdf.

As for the scrutiny of HS2, there were reports galore, from the National Audit Office, the Public Accounts Committee, the transport select committee and the House of Lords economic affairs committee. There are numerous other reviews from government and HS2 itself. They are all snapshots in time, as the scope of HS2 kept changing. The figures quoted are almost meaningless as they differed from committee to committee and report to report. The PAC report from February 2025 sums up the latest total financial uncertainty, with the scope still unknown: https://committees.parliament.uk/committee/127/public-accounts-committee/news/205518/where-now-for-hs2-pac-urges-govt-to-set-out-what-benefit-can-be-salvaged-for-taxpayer/.

Thinktanks also produced papers, such as the New Economics Foundation's *A Rail Network for Everyone* (https://neweconomics.org/uploads/files/A_Rail_Network_for_Everyone_WEB.pdf) which argued from a left-wing perspective that HS2 should be cancelled and the money reallocated to transport in the North. On the other end of the political spectrum, Andrew Gilligan's highly influential Policy Exchange report, *The Kindest Cut of All* (https://policyexchange.org.uk/publication/hs2-the-kindest-cut-of-all/) is significant in that it led to the truncation of the line under prime minister Rishi Sunak. But for a contrasting view, the Oakervee Review (https://www.gov.uk/government/publications/

oakervee-review-of-hs2), which convinced Johnson to go ahead, is also worth reading, as is Lord Berkeley's alternative report (https://tonyberkeley.co.uk/index_htm_files/rh200105%20Dissenting%20report.pdf).

As usual, the Institute for Government was a calm and thoughtful voice in the storm, offering, as in this 2021 review, *HS2: lessons for future infrastructure projects* (https://www.instituteforgovernment.org.uk/publication/hs2-lessons-future-infrastructure-projects), advice about how to avoid making similar mistakes in the future. I found their recommendation for a Commission for Public Engagement, which would involve communities in decision-making much sooner, particularly appealing.

HS2 Ltd has hundreds of webpages which demonstrate how the company ended up a behemoth, concerning itself with a whole range of activities, many of which were only tangentially linked to building a railway. A quick perusal of the legacy learning pages makes the point (https://learninglegacy.hs2.org.uk). The construction videos are the most interesting, with engineers explaining what is being built. The Colne Valley Viaduct (https://www.hs2.org.uk/building-hs2/viaducts-and-bridges/colne-valley-viaduct/), the Delta Junction (https://www.hs2.org.uk/building-hs2/viaducts-and-bridges/delta-junction/) and the Chilterns Tunnel (https://www.youtube.com/watch?v=LESdpsudU44) videos are the most worthwhile. You can also learn more about the rather absurd cut and cover tunnels being built across fields to protect rural residents from noise and preserve the landscape here – https://www.hs2.org.uk/building-hs2/tunnels/green-tunnels/greatworth-green-tunnel/ – and the extraordinary tunnel boring machines, built by the German manufacturer Herrenknecht, here: https://www.hs2.

org.uk/building-hs2/tunnels/tunnel-drives/meet-our-giant-tunnel-boring-machines/.

I haven't spent a lot of time on the technical aspects of HS2 in the book, as I was mostly interested in the political battles, but there is some extraordinary infrastructure being built – and much engineering innovation developed along the way. When the West Midlands to London part of the line is complete, not only will thousands of jobs be lost (and thousands of subcontracting firms will lose business), but the extensive knowledge base about how to build modern railways in the UK may disappear too, and with it, specialist skills.

I can also commend the former chairman Allan Cook's stocktake, which laid down some of the clearest rationale for HS2 and its potential advantages for the country in 2019: https://www.gov.uk/government/publications/hs2-ltd-chairmans-stocktake-august-2019.

Anthony King and Ivor Crewe's book *The Blunders of Our Governments* shows that HS2 is the latest in a long list of government failures in Britain. Bent Flyvbjerg's book *How Big Things Get Done: The Surprising Factors Behind Every Successful Project, from Home Renovations to Space Exploration* offers little comfort either. Mega-projects are difficult and governments around the world are bad at them.

Some of the best reporting on HS2 has been done by the *New Civil Engineer*, which I relied on as a valuable source of information. Some of their scoops – like the number of times HS2 contractors asked for more money (https://www.newcivilengineer.com/latest/hs2-contractors-have-requested-more-money-or-time-on-3000-occasions-19-12-2022/) – deserve more recognition.

Index

38 Degrees 119
51M group 119, 123
1922 Committee 188

Adonis, Andrew (Baron Adonis):
   command paper 60, 76–7
   HS2 design 72, 73–4
   HS2 rationale 4–5, 67–9, 78,
      124, 132, 133
   infrastructure renewal 62, 108,
      128, 137
Agnew, Theodore, (Baron Agnew of
   Oulton) 177
air travel:
   HS2 airport links 115, 215–16,
      231
   reductions 10, 49, 55, 58, 60,
      64–5, 132
Alderson, Edward 16
Alexander, Heidi 253
Align Joint Venture 181
Allen, David 150
Alstom 54, 243

Alto high-speed network 59
Altringham, John 146
Amersham 82, 85, 87, 122
Anderson, Joe 226
APT (advanced passenger train)
   60–1, 244
Armitt, John 130, 186
Arup:
   Euston designs 114–15, 127,
      158, 159
   wider mentions 41, 71, 92
Ashford 42, 43
Atkins 64, 101, 127
austerity policies 80, 116, 128–9, 150
Australia 58, 96
AVE (*Alta Velocidad Española*) 51,
   267
Aviation Commission 115
AVLO 51
Aylesbury 83, 85, 121, 146, 150, 180

Bacon, Richard 100
ballasted track 169, 232, 267, 271

Balls, Ed 127–9, 130, 225–6
Bank of England 37–8, 73
Basford Hall freight yard 241, 242
Bat Conservation Trust 146
bats 145–7, 177, 193, 194, 195, 263
BBC:
    HS2 coverage 80, 120, 125, 182, 194, 214, 229
    wider mentions 36, 67, 150, 169
BBV Joint Venture 181
Beeching, Dr Richard 35–6
Begg, David 126
benefit–cost ratio (BCR) 167, 191, 261
Berkeley, Anthony Gueterbock, 18th Baron 171
Bernwood Forest Bechstein's Project 146
Bethell, James 124–5
Betjeman, John 36
biodiversity 118, 142, 194–5, 263
Birmingham:
    HS2 support 129, 221–4, 256
    journey times 77, 108, 240
    wider mentions 74, 232
    see also Curzon Street, Birmingham
Birmingham Airport 13, 115, 231
Birt, John 61, 62
Blair, Tony 61, 66, 169
Blake, Judith 230
Bluebell Wood protestors 198–9, 200, 201, 202
Bombardier 243
Boote, Simon 239
Booth-Smith, Liam 213
Bostock, Mark 41–2
Bouch, Sir Thomas 25–6

Brabin, Tracy 206, 225, 230
Brackley 83, 181, 193
Bradford 230
Bradshaw, William (Baron Bradshaw) 77
Brady, Graham 188–90, 206
Branson, Richard 44–5
Brexit:
    campaign 125, 139, 154, 206
    impact 7, 122, 134, 189, 194–5, 207
    negotiations 155, 164, 166, 229
Bridgewater, 3rd Duke of 16
British Empire 7, 14, 27–8, 54, 159, 243
British Rail 34, 35–6, 42, 44, 60–1, 175
Bromberg, Andrew 159
Brontë, Charlotte 24
Brown, Gordon 4, 66–8, 72–4, 76, 132
Bruce-Lockhart, Sandy 43
Brunel, Isambard Kingdom 20–1
Brunel, Sir Marc Isambard 20
Buckinghamshire County Council 84–5, 119, 152
Bundred, Steve 99
Burke, Tom 182
Burnham, Andy:
    on HS2 cancellation 211, 213, 214–15, 231–2, 233
    HS2 support 138, 225, 229, 246
Burns, Simon 124
bus transport 35, 36, 71, 196, 214
Butler, Rob 180

Cabinet Office 102, 113, 177, 182, 255
California bullet train scheme 57–8

INDEX

Cambridge 36, 121, 135
Cambridge University 16, 73, 103, 194
Camden 89–90, 96–7, 114
Camden Council:
    construction pollution 94–5, 180
    Euston designs 158, 159–60
    homes, loss of 1, 89–90, 93–4, 253
    internal culture 92, 163
Cameron, David:
    Gillan, management of 84, 85, 87
    HS2 support 5, 79, 132, 133, 139, 154
    McLoughlin, appointment of 106–7
    wider mentions 64, 74, 113
Campaign for High-Speed Rail 123–4
Campaign to Protect Rural England 80, 142, 188
Canada 59
Canary Wharf 38, 39, 223
car industry 34–5, 46, 66, 73, 132
carbon emissions 58, 132, 136, 191, 196, 252
Carillion 166
Carlyle, Thomas 24–5
Cavenagh-Mainwaring, Edward 238
Centre for Towns 134
Challenge Panel 126
Channel 4: 79, 86, 121
Channel Tunnel Rail Link, *see* HS1
Chat Moss 16, 17
Cheshire East Council 240, 246
Chiltern Hills:
    Covid pandemic 178
    route option avoiding 133
    South Heath 82–3, 88, 117–18, 120
    tunnels 85–8, 110, 127, 151, 153, 181, 252
Chiltern Society 121
China 10, 28, 53–7, 143, 158–9
Clarendon, Earl of 20
Clegg, Nick 65
coal mining 14–15, 27, 30–1, 34, 76, 107
Colne Valley 193, 194
Colne Valley Viaduct 181, 251–2
command paper 60, 76–7
compensation:
    for cancellations 208, 219, 247
    overseas projects 55
    to public 81, 82, 94–6, 123, 153, 238, 253
    recommendations 259
    see also housing, government purchase
compulsory purchases 43, 185
concrete slab track 40, 169, 232, 271
Conservative–Liberal Democrat coalition 5–6, 79–80, 82, 128, 132
Conservative Party:
    1992 manifesto 44
    2008 conference 64
    2023 conference 2–3, 212–13
    wider mentions 83, 85, 108, 134, 188–92, 204–6
Conservative Research Department (CRD) 205
ConservativeHome 206
construction pollution 1, 9, 89, 94–5, 180, 183–4

construction safety 20, 25, 28
consultation process 81, 93, 140,
    149–53, 256, 259
contracts:
    court enforcement 195
    HS1: 42
    HS2 cancellation 104, 245, 253
    HS2 works 102, 127, 155, 166,
        172, 181–2
    legislation prior to 88, 155, 166,
        180
    recommendations 217, 260, 261
    train-building 216, 255
    train services 184
Control Risks 237
Cook, Allan 164–72, 174, 191, 197,
    231, 261
Coomes, David 194
Cooper, Yvette 128
Corbyn, Jeremy 155, 226
Corcoran, Sam 240, 245–6
cost–benefit analysis:
    Treasury 62, 76, 167, 191, 261
    wider mentions 86, 100, 110,
        116
costs:
    pre-2010: 5, 62, 63, 64, 65
    2010: 76, 77
    2012: 86–7, 101–2, 158
    2013: 101–2, 116, 127, 129
    2019: 168
    2020: 153, 173
    2022: 185, 208
    2023: 210, 216
    2024: 253
    2025: 6, 260
    Cook's stocktake 166–9, 174
    Covid pandemic 184–5

direct action 198, 199, 203, 237
environmental mitigations 146,
    194, 263
Euston 158, 160–2, 173, 208,
    216, 219
HS2 cancellation 209, 214,
    219–20, 245, 248, 249
tunnelling 87, 111, 162, 177,
    181
wider mentions 114, 115, 142,
    150, 183, 186, 188
see also compensation
Covid pandemic:
    HS2 progress 178–80, 182–5,
        186, 196, 207
    overseas projects 56
    Partygate 187, 189, 191
    rail travel 45, 180
Crane, Emma 82–3, 117–23, 125,
    126, 143, 147
Crewe:
    history of 33, 47, 243
    HS2 cancellation 232–3, 245–8
    jobs 240, 243, 244
    recommendations 264
    signalling 111, 209, 241–2, 244,
        248
Crewe HS2 Hub Draft Masterplan
    Vision 240
Crossrail:
    HS1 to HS2 link 115
    HS2 leaders from 165, 252, 253
    lessons-learned 263
    petitioning 141
    regional finance 228, 261
    wider mentions 39–41, 69, 104,
        112, 176, 267–8
Crossrail 2: 115, 161, 235, 268

INDEX

Cummings, Dominic 125, 179
Curzon Street, Birmingham 167, 222, 223, 252, 254

Defra 178, 195, 262
Delta Junction 167, 181, 252
demolition:
    Euston 1, 89–90, 93–4, 95, 157, 158
    route options 71, 82, 83
Department for Transport, *see* DfT–HS2 Ltd interrelationship
Derby, Lord 16
Deripaska, Oleg 181
design process:
    detailed design 144, 153, 166, 180, 185, 260
    Euston 114–15, 127, 158–63, 208
    McNaughton 70–2, 75
    recommendations 256–7, 260
Deutsche Bahn 31–2
DfT–HS2 Ltd interrelationship:
    joint taskforce 177, 179, 182
    processes 81, 103–4, 121–2, 123–4, 125–6, 168
    structure 5, 68, 69, 109–10
Diamond, Patrick 72–3
Dickens, Charles, *Dombey and Son* 93
direct action 119, 121, 195–203
Djemai, Nasrine 1, 89–90, 94, 97, 183–4, 253–4
Dobson, Frank 90, 94, 97
Dobson, James 200–1
Docklands Light Railway (DLR) 38, 268
Duffy, Sean P. 58

E3G 182
East Coast Main Line 45, 60, 76
economic growth 52, 56, 62–3, 72–3, 86, 100, 136; *see also* house building; jobs
economic rebalancing:
    vs. agglomeration effect 128, 138
    alternatives to HS2: 227, 232
    Brown 3, 4, 73, 132
    HS2 rationale 77–8, 80, 128–9, 132, 174, 223
    post-HS2: 224, 264
    see also levelling up; Northern Powerhouse
Eddington, Sir Rod 62–3
EKFB Joint Venture 181
electrification:
    Euston 157
    IRP 186–7
    Transpennine Route 214, 226, 230, 232
    wider mentions 3, 9, 17, 37, 107
Eliot, George, *Middlemarch* 19
Elizabeth line, *see* Crossrail
Elliff, Colin 133
Ellman, Louise 129
Environmental Act 2021: 195
Environmental Agency (EA) 195, 263
environmental factors:
    direct action 119, 121, 195–203
    environmental report 118, 120, 140–1, 142–4, 145, 153–4
    HS2 Action Alliance 119–21, 122–3, 125, 153
    Monbiot 135–7
    overseas projects 57, 143, 151
    recommendations 257, 263

wider mentions 80–1, 84, 117
see also bats; carbon emissions; trees
Essex, Earl of 20
Europe 27, 28–31, 34
European Commission 31, 143
European Union (EU) 6, 7, 51, 142, 194–5
Eurostar 6, 43, 88, 114
Euston:
- approaches 2, 167, 181, 208, 214, 216, 253
- Covid pandemic 178
- designs 114–15, 127, 158–63, 208
- direct action 197–8, 202
- history of 21, 90–1, 156–7
- homes, loss of 1, 89–90, 93–4, 253
- Oakervee report 173
- tunnels 113, 162, 181, 216, 252
- work cancellation 132, 208–9, 214, 216, 219, 252
- *see also* construction pollution

Euston Area Plan 160
Euston Development Zone 216
Euston Partnership 208, 252
Euston Strategic Board 160
Eversheds Sutherland 201
Extinction Rebellion 196–7, 202

Facebook 199, 201, 236
Fall, Kate 106
Farage, Nigel 207
Farnworth Tunnel 107
financial crisis (2008) 4, 23, 67, 72, 135, 207

financial services industry 4, 37–8, 67, 72–3, 76, 134
Financial Times (FT) 4, 127
First World War 31, 33–4
Fitzherbert, Ben 238, 239
Flinders, Captain Matthew 96
Flyvbjerg, Bent 77, 264
Forestry Commission 194
Foster, Norman 39
France 10, 24, 29–31, 151, 221; *see also* TGV (train à grande vitesse)
franchising 29, 45, 106, 184, 215, 248
Francis, John 24
Frazer, Lucy 191
freedom of information requests 122, 200
freight:
- Covid pandemic 184
- Crewe 241–2, 245, 248
- history of 33, 34, 35, 36, 37
- road 35, 37, 215, 242
- WCML 10, 132, 232, 249
Friends of the Earth 196
Fujitsu 127

Gade Valley 20
Gatwick Airport 69, 155
Germany 30–1, 44, 50
Gillan, Cheryl:
- costs scrutiny 85, 102–3, 110, 127
- mockery of 122
- Notice to Proceed 180
- petitioning 147, 149, 151, 152
- political roles 82, 83–5, 87–8, 119

INDEX

tunnels 85, 86–8, 110, 151
Gilligan, Andrew:
   cancellation rationale 186, 204–6, 209
   cancellation secrecy 210, 211, 212, 218
   as Johnson's adviser 169, 170, 178, 179–80
Gladstone, William 26, 98
Glaister, Stephen 137
Glancey, Jonathan 39
Glasgow Subway 46
Glover, Julian 107, 159
Godson, Dean 205
Golborne Link 188, 189–91, 206, 228
Gooch, Thomas 16
Gould, Georgia 163
Gove, Michael 205, 216
governance:
   Cook's stocktake 166–9, 174
   Euston 160–1
   hybrid bills 152–4, 262
   Monbiot's views on 136
   Network Rail 104, 111, 153, 167, 190, 252
   Number 10: 175, 177, 179, 189, 212
   PAC 2025 report 251
   recommendations 255–63
   top-down 10–11, 81, 149, 187
   see also DfT–HS2 Ltd interrelationship
Grand Central Railway 243
Grand Junction Railway 33, 243
Grayling, Chris 168
Great Missenden 152

Great Western Railway (GWR) 21, 26, 40, 184
Greater Manchester Combined Authority 138, 227
green bridges 145, 193–4
Green Party 136, 144
'green path' outdoor walkway 114–15
green transport 53, 57, 64–5, 66, 80, 144
green tunnels 86–7
Greengauge 21: 126, 248
Greening, Justine 81, 85–6, 113, 257
Greensill Capital 113
Greensill, Lex 112–13
Grimshaw 158, 159, 240
Grossman, David 125
Guardian 64, 99, 136, 190
Gunn, Simon 34, 35

Hammond, Philip 79, 80–1, 83
Handsacre Junction 181, 214, 222, 248–9
Hanham, Joan (Baroness Hanham) 77
Hansard Society 105, 140, 147, 148
Hardwick, Philip 156
Harper, Mark 207–8, 210
Harrison, Trudy 191
Hayward, Sarah 90, 93–4, 158
Heathrow:
   Crossrail 40, 115
   HS2 spur 81, 86, 88, 115, 119
   third runway 64–5, 152, 216, 231
Heathrow Hub 119
Hendy, Peter 171, 190, 208, 252
Hennessy, Peter 98
Heseltine, Michael 41, 211

Heywood, Jeremy 112–13
Higgins, Sir David 109–10, 115, 155, 165–6, 232
high-speed lite solution 232
High-Speed North 173
High Speed (Preparation) Bill 109, 127, 130
High Speed Rail (Crewe to Manchester) Bill 141, 152, 188, 189–91, 224
High-Speed Rail Group 191, 210, 213, 232, 248, 255
High Speed Rail (London to West Midlands) Bill:
    environmental report 118, 120, 142–4, 145, 153–4
    Euston 114, 158, 162
    petitioning 141, 144, 147–9, 151–2, 225, 262
    wider mentions 109, 140–2, 168–9, 224
High Speed Rail (West Midlands to Crewe) Bill 141, 152, 211, 224, 232, 246
Highways England (National Highways) 153
Hitachi Alstom 183, 216, 254
Hodge, Margaret 85, 98–104
Holden, Rob 43
Hoon, Geoff 68, 70, 73
Hooper, Dan ('Swampy') 197–8
Hooper, Rory 197, 202
Hopkins Architects 39
Houchen, Ben 206
house building:
    Chiltern Hills 85, 148
    Euston 214, 216

    North 75, 224, 230, 240
housing:
    affordability 49, 52
    blighting of 94–6, 149, 180, 183–4, 235, 236–8
    demolition 1, 89–90, 93–4, 253
    government purchase 43, 95–6, 185, 235, 236–8
    recommendations 264
    route options 5, 71, 82, 83, 149–50
HS1:
    franchising 215, 248
    governance 93
    HS2, link to 86, 88, 92, 113–15, 153, 236
    lessons-learned 101–2, 263
    wider mentions 6, 41–4, 69, 141, 259, 268–9
HS2 Action Alliance 119–21, 122–3, 125
HS2 cancellation:
    Burnham's response 211, 213, 214–15, 231–2, 233
    Crewe 232–3, 245–8
    Euston 208–9, 214, 216, 219
    rationale 186, 204–7, 209, 211, 220, 247
    secrecy 210–13, 218, 231, 261
    Street's response 138, 210–13, 231, 233
    Sunak's announcement 2–3, 213–14, 233
    transport select committee 104
HS2 design board 160
HS2 Extinction 201
HS2 Ltd, *see* DfT–HS2 Ltd interrelationship

INDEX

HS3: 68–9, 129, 226–7, 269
Huddersfield Station 22, 25
Hudson, George 24–5
Hunt, Jeremy 210
Huskisson, William 17–18
hybrid bill select committees 141, 147–9, 150, 262
hybrid bills 88, 105, 107, 152–4, 185, 224

ICE (Intercity Express) 50, 269–70
India 27–8, 54
Indonesia 56–7
industrial strategy:
    national 75, 78, 133, 134–5, 168
    regional 210, 247
Ineco 52, 270
Infrastructure and Projects Authority 104, 177, 255
Institute for Government 137, 151
Institute for Public Policy Research (IPPR) 182
Integrated Rail Plan (IRP) 186, 230
InterCity 125: 60, 61, 131
Ireland 23, 26

Jacobs 245
Japan 10, 47–9, 50
Javelin train 6, 43, 73
Javid, Savid 173, 179, 191
jobs:
    Crewe 240, 243, 244
    growth 75, 124–5, 132, 217, 223, 224
    HS2 construction 100, 167, 216, 251
    recommendations 264
Johnson, Boris:
    appointments by 169–71, 175, 192
    Covid pandemic 178–9, 182, 185, 187
    HS2 approval 173
    no-confidence votes 187, 189, 190
    project cancellations 206, 229
    resignation 191
    wider mentions 41, 73, 111, 158, 194, 211
Johnson, Stanley 95–6
joint taskforce 177–9, 182
Jones' Hill Wood 197, 201
Joyce, David 94
Jubilee Line Extension (JLE) 38–9, 40, 159
judicial reviews 119, 122, 153, 196
Just Stop Oil 202, 203

Kelly, Bernadette 218–19, 253
Kelly, David 169
Kelly, Gavin 74
Kelly, Ruth 63, 65, 67–8
Kemble, Fanny 17–18
Kent 42, 43–4, 50
Khan, Sadiq 41, 63
Kier 181, 201
Kilsby Tunnel 20
Kingdom, Sophia 20
King's Cross station 44, 91, 92
Kirby, Simon 110

Labour Party:
    2005 manifesto 4, 61
    2008 conference 67–8
    2013 conference 129
    2023 conference 246

wider mentions 2, 9, 63, 127–30, 225–7, 229
Laird, Philip 58
land:
    19th century 15–16, 20
    government purchase 235, 238–9, 248, 249
Leadsom, Andrea 83
Leeds 77, 128, 162, 214, 224, 229–31
Leeds City Council 206, 230
Leeds/eastern leg 185–6, 187, 206
Leese, Richard 75, 227–8
Lendlease 161, 208
levelling up 73, 132, 191, 244
    abandonment of 3, 137, 192, 205, 211
Lewis, Mary-Ann 159
Liberal Democrat Party 65
Lidington, David 83, 122
Little Missenden 87, 88
Liverpool 128, 186, 226, 231, 246, 264
Livingstone, Ken 41, 170
London:
    City of (financial services) 4, 37–8, 67, 72–3, 76, 134
    HS2 terminus options 64, 91–2, 97, 173, 215
    inward-looking 10, 134–5
    mayor of 160, 163, 170, 228, 253, 261
London & Continental Railways (LCR) 42, 44, 69
London and Northwestern Railway (LNWR) 243
London Midland & Scottish Railway 93
London Overground 21, 114
London to Birmingham line 19–20, 156
London Transport 38
London Underground:
    HS1, link to 44
    HS2 station plans 91, 115, 216
    wider mentions 46, 69, 163, 203
    see also Jubilee Line Extension (JLE)
Lucas, Caroline 144

MacAdam, Ailie 42–3, 93
McBride, Damian 68
Macdonald, Quentin 133
McLoughlin, Patrick, (Baron McLoughlin) 106–10, 116, 134, 143, 153, 154
McMahon, Jim 187
McNaughton, Andrew:
    design process 70–2, 75, 78
    Euston 161
    HS2 cancellation 249
    mitigations 87, 111, 145–6, 148
    route 71, 77, 91, 111, 148
Macpherson, Nicholas 69
Madeley and Whitmore Villages Stop HS2 group 236
Maier, Jürgen 212
Major, John 39, 75
Major Projects Authority 102, 104
Mallender, Deborah 234, 235, 236–9
Manchester:
    2023 Conservative Party conference 212, 215
    HS2 cancellation 224, 227–9, 232–3, 264

INDEX

HS2 support 129
journey times 77, 128
see also Burnham, Andy
Manchester Airport 13, 115, 231
Manchester City Council 75, 206, 227
Manchester Piccadilly 167, 206, 228, 231, 254
Manchester to Liverpool line 15–18
Mandelson, Peter 127
manufacturing industry:
   history of 15, 27, 34, 47, 75, 243
   HS2: 126
   revival policy 73, 75, 134, 225
Marples, Ernest 35
May, Theresa 154, 155, 164, 169, 189, 229
Merriman, Huw 187, 247
Metronet 69
Mexborough 185
Midlands Engine Rail 230
Midlands Rail Hub 171, 214
Miliband, Ed 103, 128, 185, 226
Miller, Maria 85
mini-Shinkansen 50, 270
Minton, Anna 124
Mirza, Munira 170
Mitchell, Andrew 213
modular design 38–9, 50, 54, 217, 258, 264
Monaghan, Ross 200–1
Monbiot, George 135–7
Morgan, Sadie 160
Morgan, Terry 165
Morse, Amyas 99
Mullan, Kieran 246
Munro, Alison 70, 71

Murison, Henri 75, 214, 231
Murphy, Luke 182
Murty, Akshata 207

Nandy, Lisa 134
National Audit Office (NAO) 99, 104, 116, 153, 166, 178, 208
National Infrastructure Commission 104, 130, 137, 185–6, 205
National Rail 46, 184
National Trust 80, 146–7, 149–50
nationalisation 30–1, 34, 184, 226, 248
Natural England 145–6, 147, 178, 263
Nesbit, E., *The Railway Children* 33
Network Rail:
   Crewe signalling 209
   Euston 97, 157, 161, 162, 163, 245
   Farnworth Tunnel 107
   formation of 44
   HS2 governance 104, 111, 153, 167, 190, 252
   HS2 leaders from 70, 71, 109, 161, 171
Nichols, Connor 201–2
noise pollution 87, 94–5, 149, 183, 235
Norman, Jesse 177
Northamptonshire 20, 83, 133, 193
Northern Powerhouse 3, 73, 225, 233
Northern Powerhouse Partnership 75, 214, 231
Northern Powerhouse Rail 171, 176, 186, 187, 214, 230–1, 269

Northern Rock 66
Northolt Tunnel 111, 181
Notice to Proceed 169, 172–3, 178, 179–81, 182
Nottingham 129
Nottinghamshire 9, 230
Number 10:
    HS2 governance 175, 177, 179, 189, 212
    HS2 opposition 170
    HS2 proposal 73–4
    leadership challenges 210

Oakervee, Doug 171–3
Oakervee report 171–3, 176, 205, 208
Obama, President Barack 119–20
Occupy protests 197
Old Oak Common:
    alternative terminus 97, 173, 215
    Euston link 113, 252
    hub 115, 254
    route options 113, 115, 133
Oldfield, Leah 201
Olympia and York 39
Olympic Games 5, 6, 42, 48, 130, 260
Osborne, George:
    austerity policies 80, 128–9, 150
    Northern Powerhouse 73, 225
    West Kowloon MTR station 158, 159, 257
    wider mentions 111–12, 130, 133, 139, 154, 211
Oxford 36, 121, 135
Oxford University 148

Packham, Chris 196
Paddington station 21, 40, 92

Palin, Michael 157
Paoletti, Roland 39
Park, Richard 232
Parker, Richard 213
Partygate 187, 189, 191
passenger journeys:
    Covid pandemic 184
    demand forecasts 100, 101–2, 233, 248
    overseas projects 48, 52
    statistics 33, 46, 107, 112
passenger safety 17–18, 25–6, 44
Pease, Edward 15
Pendolino trains 45, 61, 244, 254–5
petitioning 141, 144, 147–9, 151–2, 225, 262
Peveril 245
Pincher, Chris 191
Policy Exchange 204–5, 209
Prentice, Gordon 176
Prentis, Victoria 122
Prescott, John 42
Pritchett, James Pigott 22
privatisation 44–5, 48, 107
Prout, David 109
Prussia 29, 30
Public Accounts Committee (PAC) 85, 98–104, 178, 219
    2025 report 251, 253, 254
Public Order Act 2023: 203

Raab, Dominic 180
Rail Freight Group 171, 191
Railtrack 44
Railway Heritage Centre 244–5, 248
Railway Industry Association 191

INDEX

Reeves, Rachel 232, 263
Reform Party 207
Regent's Park Estate 89–90, 253
Regulation of Railways Act 1889: 26
road pricing 10, 62, 63, 65
roads:
   congestion 61, 63, 215, 232
   environmental legislation 57
   freight 35, 37, 215, 242
   history of 15, 21–2, 34–5
Rocket 17–18
Ronay, Barney 157
Roosevelt, President Theodore 28
Rotherham, Steve 186, 231, 246
Rothschild family 150
route options:
   Adonis 73–4, 76
   housing 5, 71, 82, 83, 149–50
   London terminus 64, 91–2, 97, 173, 215
   McNaughton 71, 77, 91, 110–11, 148
   Old Oak Common 113, 115, 133
   preparation 153, 166, 180, 260
   recommendations 172, 260, 262
Rowlands, Jenny 163
Rowlands, Sir David 69–70
Royal Automobile Club (RAC) 66
RSPB 146–7
Rukin, Joe 119
Ruskin, John 136
Russia 31, 236
Rutnam, Philip 100, 102, 103, 109, 154–5

St James' Gardens Burial Ground 96
St Pancras International:
   Euston link 113, 114, 115
   HS1: 6, 41, 42, 44, 73
   route options 64
Sankey Viaduct 16–17
Scotland:
   cross-border travel 9, 120
   HS2: 65, 67, 81, 138, 217
   road tolls 63
SCS Railways 181
Second World War 31, 34
Sefton, Lord 16
Shapps, Grant 170, 176, 186, 189–90
Sheephouse Wood 145
Sheffield 77, 129, 185, 228, 229–30
Shinkansen 47–9, 50, 109, 255, 259, 271
   tracks 48, 49, 109, 260, 271
signalling:
   Crewe 111, 209, 241–2, 244, 248
   history of 26, 37, 44
   HS1: 43
   HS2: 70, 133
   Shinkansen 48
slab track 40, 169, 232, 271
Smallwood, Nick 177
Smiles, Samuel 14, 20
Smith, Greg 180
Solihull 231, 252
South Heath 82–3, 88, 117–18, 120
South Yorkshire 86
Spain 50–3, 143, 221, 228, 256, 259
Speight, Becky 144
Starmer, Keir 94
state ownership 29–34, 45, 52, 184
steam locomotives 14–15, 17–18, 21, 36, 54, 243
Steer, Jim 126

Stephenson, Andrew:
   HS2 Minister 170, 175–80, 185, 190, 191, 215
   IRP 186–7
   wider mentions 182, 192, 195
Stephenson, George 14–15, 16, 20, 156
Stephenson, Robert 13–14, 15, 17, 19–20, 156
Stewart, Iain 210
Stoke 234, 236
Stonehenge Heritage Action Group 201, 202
STRABAG 181, 236
Stratford International 38, 40, 41, 43, 113–14, 115
Straw, Jack 129
Street, Andy:
   on HS2 cancellation 138, 210–13, 231, 233
   HS2 support 171, 221–4, 256
   Johnson's HS2 review 170, 171
Sunak, Rishi:
   cancellation announcement 2–3, 213–14, 233
   cancellation rationale 207, 209, 211, 220, 247
   cancellation repercussions 138, 216–17, 235
   cancellation secrecy 210, 212, 213, 218, 261
   Prime Minister, appointed as 192, 204
   wider mentions 184, 191, 202, 205
Swynnerton 198, 200, 238, 239

TaxPayers' Alliance 121, 178, 183
Taylor, James 201
Telford, Thomas 107
Temple 71
Tett, Martin 119
TfL (Transport for London):
   Crossrail 40, 261
   HS1: 44
   station proposals 91, 97, 104, 160, 216
   see also London Underground
TGV (train à grande vitesse) 41, 43, 49–50, 51, 109, 271
Thames Tunnel 20–1
Tharoor, Shashi 28
Thatcher, Margaret 39, 44, 61, 75, 106, 241
Thompson, Jon 174, 209
Thurston, Mark 168, 209
The Times 23, 68, 127, 136, 211
Tjia, Linda Yin-nor 54
Tomboy Films 123
Toton 247
tracks:
   gauge 21, 48, 270
   maintenance 44, 104
   Shinkansen 48, 49, 109, 260, 271
   slab *vs.* ballasted 40, 169, 232, 267, 271
   TGV 49–50, 109, 271
   WCML 60
Trafalgar Entertainment Group 240–1
train-building contracts 183, 216, 255
tram systems 187, 214, 223
Transpennine Route 25, 176, 187, 230, 232
Transport and Works Act 1992: 152
Transport for London, *see* TfL (Transport for London)

INDEX

transport select committee 104, 129, 151
transport strategy:
  national 8, 9–10, 22, 35, 61–6, 137
  regional 8, 137, 185, 214, 230
Travers, Tony 99, 171
Treasury:
  cost–benefit analysis 62, 76, 167, 191, 261
  financial crisis (2008) 72, 73
  high-speed rail 4, 65
  HS2 cancellation 208, 210, 219, 220
  HS2 opposition 69, 77, 111–12, 172
  joint taskforce 177, 178, 182
  pre-HS2 projects 39, 42, 45, 112
  recommendations 248, 258, 261
trees 144–5, 163, 170, 194, 197, 200
Truss, Liz 191–2, 204
tunnels:
  19th century 20–1, 25
  for bats 147, 177, 193, 195, 263
  Chiltern Hills 85–8, 110, 127, 151, 153, 181, 252
  Crewe 241, 248
  direct action 197–8, 199–203
  Euston 113, 162, 181, 216, 252
  HS2 overview 82, 145, 169, 233, 257, 263
  overseas projects 48, 50, 52
  pre-HS2 projects 40, 41–2
Turner, J. M. W. 21, 26
Turweston 193

UK Independence Party (UKIP) 134
Ukraine war 7, 31, 192, 204
ultra low emission zone (ULEZ) 63
Union Railways 42
United States of America (USA) 57–8

viaducts:
  HS2: 150, 153, 167, 181, 193, 251–2
  wider mentions 14, 16–17, 42, 48, 54
Vick, David 148
Villiers, Theresa 64, 68
Virgin 45, 61
Vivis, Mark 85

Waddesden 148, 150
Wales 36, 83, 85, 87, 138, 241
Walker, David 99
Walley, Joan 235
Walmsley, Ian 139
Wellcome Collection 96
Wellington, 1st Duke of 17–18
Wendover 197, 201
West Coast Main Line (WCML):
  APTs 45, 60, 61
  capacity 10, 86, 108, 132, 215, 232–3, 242
  construction 19
  high-speed lite solution 232–3
  HS2 route options 71, 76, 92, 133
West Kowloon MTR station 158–9, 257
West Midlands 74, 213, 222, 233
West Yorkshire 187, 206, 230
Westbourne Communications 123–4
Western Slopes project 194

Westminster 38, 39, 92
Weston, Bruce 117, 120, 123
Wharf, Hilary 117, 120, 123
Whitmore 234–9
Whoosh 56–7
Wild, Mark 180–1, 252–3
Wildlife Trusts 144
Wilmslow East 244
Wolmar, Christian 23, 80, 131–3, 134–5, 139
Woodhead Tunnel 25
Woodland Trust 142, 144, 146

Wright, Jeremy 83
Wright reforms 99
WS Atkins 165

'Yes to High-Speed Rail' campaign 123–5
Yin Zihong 55, 56
YouGov 138, 227
YouTube 125

Zola, Émile, *La Bête Humaine* 30